MAN IN BIOSPHERE

MAN IN BIOSPHERE

A Case Study of Pachmarhi Biosphere Reserve

Editor
SADHAN SENGUPTA

GYAN PUBLISHING HOUSE
NEW DELHI-110002

Man in Biosphere A Case Study of Pachmarhi Biosphere Reserve

Rs. 540

ISBN: 81-212-0944-7

Published in 2007 in India by
Gyan Publishing House
5, Ansari Road, Darya Ganj,
New Delhi-110002
Phones : 23282060, 23261060
Fax : 011-23285914
E-mail: books@gyanbooks.com
Website: www.gyanbooks.com

Lasser Type Setting by: Rudra Computer Graphics, Delhi
Printed at : Salasar Imaging System, Delhi

Contents

List of Tables

List of Maps

Foreword

The present volume entitled *Man in Biosphere – A case study of Pachmarhi Biosphere Reserve* is the outcome of the research project of the Anthropological Survey of India undertaken in the Tenth Five Year Plan Project – *Cultural Dimension of Tourism*, reoriented as *Man in Biosphere*. Pachmarhi is situated in Madhya Pradesh covering the hills of Satpura and falls into three districts i.e. Hosangabad, Betul and Chhindwara. It was notified by the Govt. of India in the month of March, 1999. The total area of Pachmarhi Biosphere Reserve is 4987 sq.kms. and it covers three wildlife conservation units i.e. Bori Sanctuary 485.72 sq.kms; Satpura National Park 524.37 sq.kms. and Pachmarhi Sanctuary 417.78 sq.kms. The core zone of Pachmarhi Biosphere Reserve is 524.37 sq.kms. which is also known Satpura National Park. The remaining area of 4462.93 sq.kms. serves as a buffer zone.

Pachmarhi Biosphere Reserve falls under the Mahadeo hill and is visited by large number of tourists. The altitude varies from 322 to 1352 mtrs. The general configuration area is hilly, undulating terrain and has some deep gorges around Mahadeo Hills. The Pachmarhi hill of Madhya Pradesh is also known as queen of Satpura. The research work under this project, mainly dealt with issues of indigenous knowledge of resource management, traditional management of tourism, water shed management, ecology and material culture, ethnic composition of buffer zone and geographic and demographic aspects of Pachmarhi Biosphere Reserve.

I am quite sure, this volume is going to serve an important purpose in bridging the gap between biodiversity and cultural diversity and understanding the role of man in conservation and development of the environment. It also focuses on how traditional tourism has been taken over by modern tourism and the local inhabitant of the area lost their livelihood. The settlement issues is also important because the tribal population of the core villages are willing to move out, if proper rehabilitation is provided. I congratulate the team of Shri Sadhan Sengupta for completing this research work.

(V.R. Rao)
Director-in-Charge

Preface

Concern for forest and its surrounding environment is as old as the human civilization. Forest has been the abode of human population for sustenance by utilizing the forest resource and at the same time protecting it through traditional form of forest management. Thus there is a symbiotic relationship between the forest ecosystem and the human population. This symbiotic relationship got strained after the industrial revolution when forest was looked upon as a source of wealth. This attitude changed the total utilitarian pattern and from this period forest degradation started resulting in vanishing of the forest cover from many parts of the earth surface. India was no exception to this change and with the coming of the colonial power forest became the prime target for massive exploitations. Laws were enacted to take full control of the forest resources and reallocation / resettlement of human population started during this period. The so called autochthonous population who were residing in the forest had to bear the brunt very heavily. The village settlements inside forests were relocated at different places where they were under the full control of the forest department for supplying labour in the day to day activity of the department. Baiga Chack, in Madhya Pradesh, is a good example where the villagers were shifted from the dense forest areas to the valley plains resulting changes in land utilization and livelihood pattern. No longer shifting cultivation was allowed by the governance. At a later stage when forest were more exploited, different stake holders started making inroads for timber exploitation, collection and sale of various types of forest produce. Encroachment of forest land started on a massive scale. Forest cover was cleared to accommodate agriculture, mining, industry, and other human induced activities. Whatever pristine ecosystem was left became the target of eco-tourism in the recent years followed by construction of various types of infrastructural support system for the growing tourism industry. People, who were living in this forest ecosystem, became a silent spectator losing all their rights and privileges in the forest ecology.

During the mid 18th century man started thinking loudly regarding the problems of forest conservation and its protection. At the same time ways and means were looked for how to restore the forest ecology so that the degradation of environment is arrested. The journey from Stockholm to Rio de Janeiro forced the civilized society to launch for UNESCO's Man and Biosphere (MAB) programme in 1970. The basic objective of this programme is to minimize man-nature conflict and to protect and conserve the biosphere wherever it is possible in its pristine ecosystem. It also suggested the sustainable development of human population living in and around biospheres and tried to establish a linkage between biodiversity and cultural diversity. Anthropological Survey of India in it's 10th Five year Plan project undertook a detail study on the project 'Cultural dimensions of tourism' which was reoriented as 'Man in Biosphere' project. Under this project the Survey decided to study first three biospheres namely Pachmarhi BR, Nilgiri BR and Sundarban BR out of the fourteen Biosphere Reserves declared by the Govt. of India till date. The issues identified for studies in these biospheres are mainly anthropogenic threats, ecotourism, traditional forest management, resource utilization and sustainable development, rehabilitation issues, etc.

The volume entitled ***Man in Biosphere: A Case Study of Pachmarhi Biosphere Reserve*** presents a detail account of man's role in changing the forest eco-system. Pachmarhi Biosphere Reserve is situated in the Satpura hill ranges of Madhya Pradesh in the districts of Hosangabad, Betul and Chhindwara. Total area of the PBR is 5987 sq.km. in which the core zone consists of 524-37 sq.kms. The buffer zone has an area of 4462.93 sq.km. Pachmarhi Br has three wildlife conservation units, Bori Sanctuary, Satpura National Park and Pachmarhi Sanctuary. Among them the Satpura National Park has been year marked as the core zone of PBR. Pachmarhi BR is primarily inhabited by three major tribal groups namely Gond, Mawasi and Kurku along with cast populations. PBR has a history of blatant destruction of bio-diversity due to exploitation of the sal and teak forest. The core zone of the PBR has 47 villages in its location according to the latest information given by EPCO, Bhopal.

The volume consists of seven articles which are based on detailed fieldwork by the researchers of Central regional Centre of Anthropological Survey of India, located at Nagpur. Apart from a preamble by Sadhan Sengupta, the first article entitled PBR : Core area Study by S. Sengupta, D.N. Pandey and A. Roy. In this article, a detailed account of the PBR core area has been presented along with the problems faced by the villagers which are inhabiting the core area. It has noted that the villagers with their own effort has developed their agriculture with the help of irrigation wherever it has been possible. Due to constraint, the villagers are willing to move out if proper rehabilitation is provided. The second article by Samit Ghosal on Management of Natural Resource and Practice of Indigenous Knowledge System of the local people in the core area deals with reference to agriculture and animal husbandry. Wherever land is available people have tried to develop agriculture with the help of modern infrastructure like putting a pump set for irrigation facilities or constructing a check damp to store the water. Surajit Kumar Pal in his article The Watershed in the Core Zone villages of PBR presents a good account of the water utilization system of the people of the core zone. Water is scare in the Pachmarhi hills but the local people have ways and means to conserve the water for irrigation and domestic purposes. Ecology and Material Culture of PBR by Tamal Maity has given a good account of the material culture of the people of Pachmarhi. The people of Pachmarhi have utilized the local resources only for their material culture. It is understood that population pressure generally comes from the buffer zone so it has been decided to deal with this problem of demography and ethnic composition of the buffer zone of PBR. In this article Mukesh Sahay and Sheela Dasgupta has presented a comprehensive account of the demographic profile of PBR based on census report. The last article by R.B. Thakre gives us an account of the geographic and demographic aspects of PBR based on the maps supplied by EPCO, Bhopal. With the help of this map one can gets a detailed account not only of demography but also of the physical condition of PBR. The volume has information on the people of PBR and their interaction with the forest eco-system for their livelihood. They have taken care to see that the forest ecology is protected so that they should not face any problem. But on external forces they do not have any control and due to this destruction of forest environment is going on regularly.

Sadhan Sengupta
Core Member,
Pachmarhi BR Project

Introduction

Sadha Sengupta

Introduction

(a) Cultural Anthropology and Tourism: A Review

During the last seven decades of the twentieth century, tourism has developed as an industry throughout the world and contributing to the national income of many countries in a significant way. In simple words, the phenomenon of tourism may be described as a temporary form of nomadism in which individuals come out of routine day-to-day life along with their respective socio-cultural environments and mix with other local people on their native places in distant places. As it offers culture contacts among different groups of people, tourism has attracted the attention of cultural anthropologists since 1970 (Van Den Berghe-1998, 551-552). A few important studies on this subject have been conducted by Turner and Ash (1976), Mac Cannel (1989), Smith (1989) and Nash and Smith (1991). Review of these literatures reveals the following major observations:

1. Tourism implies transience. It is a form of temporary nomadism.
2. Tourism is characterized by encounters between strangers who do not expect a long lasting relationship and whose transactions tend to be instrumental, limited in their aim, not repeated, hence open to mutual attempts at manipulation for short term gains.

3. Tourists and local people are almost always quite different from one another in culture, language, religion, social class and a variety of other social and cultural characteristics.
4. These differences make the local people of a geographical space unique to the tourists.
5. Tourism in the form of ethnic relationship where communication is frequently impeded or truncated by language barriers and multiple forms of misunderstandings, breeches normal rules of interaction. It results in many mutual stereotypes, irritations and low expectations along with occasional pleasures and amusements in unfamiliar and unexpected situations.
6. Tourists and local people of a tourist spot often tend to make a spectacle of themselves to each other in many ways, which are more self-conscious than normal day-to-day interaction.
7. Interactions between tourists and the local people are not only qualitatively truncated but also spatially segregated in most cases. Majority of the tourists usually concentrate in a limited number of tourist spots and in a limited range of specialized facilities, i.e., hotels, eating places, travel agencies, and means of transport, and so on. Only a very few tourists dare to penetrate beyond this tourist front stage.
8. According to Turner and Ash (ibid), majority of tourists come from the rich countries or from the privileged classes of the poorer countries. Tourists wealth and high status put them in a relatively advantageous position which are often resented by the local people in a place of tourist interest. Otherwise, the local people can take advantage of tourists through their much greater knowledge of local conditions, prices, services, and so on.
9. Many tourists desire a change in their normal routine life and gain new experiences of an unknown place. But their individual interests differs a lot due to their individual preferences on matters such as climate, flora, fauna, architecture, archeology, cuisine, music, local festival, places of worship, etc.
10. The tourist's quest for original and authentic souvenirs or performing arts compels the local people to be careful about

their productions, behaviors etc., and this consciousness helps to strengthen their traditional cultural identities or even culture change.

(b) Background of the Project

Tourism is growing very fast as an industry in India depending on its own qualities. Government of India is also taking much interest for its propagation through the Ministry of Tourism. The Ministry of Tourism along with its state counterparts have their own priorities, such as promoting interesting packages to both domestic and international tourists covering all economic groups; conservation of tourist spots; improvement of infrastructure for transport, communication, accommodation, safety and security of the tourists, healthcare and sanitation facilities, entertainment, and so on. They also appreciate information and suggestions on financial as well as structural management of resources, which are available in a place of tourists' interest. As mentioned above, the cultural anthropologists have their own priorities.

In the 10th Five Year Plan, Anthropological Survey of India undertook a national project entitled 'Cultural Dimension of Tourism in India'. Accordingly field investigations were conducted at various tourist places and the findings were presented in a workshop. The reports mainly described the tourist places and a partial view of the tourists who visited those places. They did not contain any suggestions in which Ministry of Tourism could be interested nor any sound analysis of data from which a new theory could be initiated. This situation could be equated with the event for the propagation of 'industrial forestry' in Bastar, Madhya Pradesh (Anderson and Huber-1988:66).

Partial modification of the said project was done in July 2004 and it was decided by the competent authority that, henceforth, Anthropological Survey of India would study the impact of 'eco-tourism' in various biosphere reserves in India. But the main title, i.e. 'Cultural Dimension of Tourism', which was duly approved by the Ministry of Culture, Government of India, remained unchanged. In the meeting, it was also decided that a team would undertake a thorough investigation of the Pachmarhi Biosphere Reserve in

Madhya Pradesh and Shri S. Sengupta would act as the Core Member i.e. Leader-cum-overall In-charge of the project. It was also decided that before undertaking actual field investigation, the nodal agency of each biosphere reserve should be contacted to gather basic information on the study area and then formulate the actual research design.

The Environmental Planning and Co-ordination Organisation (EPCO), Bhopal, Madhya Pradesh looks after the affairs of the Pachmarhi Biosphere Reserve (PBR). The Core Member along with Shri D.N. Pandey, Human Ecologist, C.R.C., Nagpur visited EPCO in September 2004. It was found that PBR's affairs are being managed by two different organizations. The EPCO mainly does planning and monitoring researches in the PBR along with releasing of funds for different activities; on the other hand, the State Forest Department and the Satpura Tiger Reserve Project authority are entrusted for the implementation and monitoring of the various research and developmental activities throughout the PBR territory. The EPCO has systematically covered flora, fauna, topography, land use pattern, climate, water bodies, etc., within the PBR with the help of different organizations and published the data. But the human component of the biosphere has remained unchanged. After collecting necessary information from EPCO, the researchers contacted the State Forest Headquarters in Bhopal. State Forest authority simply advised them to drop the research plan, as they did not allow 'eco-tourism' activities inside the PBR. Moreover, it was told that the forest department was trying to rehabilitate all the villagers living inside the core area of PBR (i.e. the Satpura National Park) somewhere else in the periphery of PBR. A long discussion took place on this issue. Views of the State Forest Department in support of their plan in the PBR are given below:

1. Alike human beings, the flora and fauna too have equal rights to exist in a given geographical area. Otherwise, ecological balance would be disturbed and would affect the biodiversity of the area adversely.
2. In an area, human population, due to various factors grows faster than that of the native flora and fauna. Therefore, there exists an inverse relationship between them.

3. For the future prosperity of our country, strict conservation of the 'gene pool' in the 'core area' of the biosphere reserves is an absolute necessity.
4. Loss of a single species causes a lot of disturbances to other related species of flora and fauna and finally causes an immense loss to the whole ecology.
5. 'Gene Pool' areas must be saved from any sort of human manipulation.
6. Forest villages inside PBR core area have no facilities like safe drinking water, proper roads, transport, primary health centres, and so on. These facilities can only be provided to them if they are shifted to the periphery of the PBR (i.e. contact zones).
7. Government of Madhya Pradesh has already taken decision to remove all the human settlements from the PBR. Shifting of Dhain village is going on and the others are in the pipeline. Arrangements for adequate compensations are being done so that the displaced villagers may be able to start a fresh life in a new situation without much difficulty.
8. It was also told that, to assess the socio-economic condition of the PBR villages, Department of Forest and Environment, Madhya Pradesh, is undertaking village surveys with the help of the Centre for Rural Development and Environment, Bhopal (an NGO organization). They have already completed survey in 30 villages in Hoshangabad block. At present, surveys are also being undertaken in 30 villages of the Pipariya block and 40 villages in the Chhindwara block. Altogether these villages constitute roughly about 20% of the total villages in the PBR.

From the above discussion, the researchers got some clue about how to proceed for the next step abandoning the idea of studying impact of eco-tourism. Instead the issue related to the necessity of conservation of bio-diversity in a geographical space vis-à-vis conservation of traditional rights of the indigenous people over their land seemed to be more appropriate. Mentioning of traditional rights of the indigenous people over their land and forest could have been disastrous in front of the forest officials. So, to save the situation, the researcher desired to study the present socio-economic condition of the core area villages of PBR to which the forest department reluctantly agreed.

(c) Objectives

The objectives of the present study, therefore, were revised and keeping aforesaid points in the mind, the following issues were taken up:

1. To investigate current socio-economic conditions which prevail in the core villages of PBR.
2. To determine the effects of the declaration of biosphere reserve on their corporate socio-economic as well as religious life.
3. Whether their present economic activities have any adverse impact on the conservation activities of biodiversity and natural resources.
4. To find out possible ways so that an amicable solution for conservation of biodiversity, promotion of tourism and undoing of difficulties faced by the local people of the core area of PBR could be obtained.

The above mentioned objectives will be further elaborated in the remaining chapters of this report to reveal the complexity of the aforesaid problematic situation.

(d) Pachmarhi Biosphere Reserve: A brief review

Pachmarhi is a reputed hill station of Madhya Pradesh. Due to rich biodiversity and unique ecosystem of its adjoining regions, the Government of India has designated Pachmarhi as 'Biosphere Reserve' in 1999 (vide notification No. J-220116/17/94, dated 03.03.1999).

Protected area of the Pachmarhi Biosphere Reserve (PBR) lies between 22°11' to 22°56' N latitude and 77°47' to 78°52' E longitude and partially covers three districts of Hoshangabad, Betul and Chhindwara of Madhya Pradesh. Total area of PBR is 4987.38 Km^2 and it envelops three separate wildlife conservation units, viz. Bori Sanctuary (485.72 Km^2), Satpura National Park (524.37 Km^2) and Pachmarhi Sanctuary 417.78 Km^2). Among them, Satpura National Park has been marked as the core zone of PBR. The remaining area of 4462.93 Km^2 serve as the buffer zone. All the above stated wildlife conservation units together were declared as Tiger Project Area in 2000 (Singh, 2003:103-105).

The boundary of the Satpura National Park which is the core area of Pachmarhi Biosphere Reserve (PBR) where the present study was undertaken from February 02, 2005 to March 10, 2005 is given below.

North: From Tawa Dam to Denwa River in the northeast

East: Eastern boundary of forest division No. 485; southern part of forest division No. 236, 257, 266; eastern boundary of forest division No. 240, 266 and 267.

South: District boundary of Chhindwara district

West: Northern boundary of Tawa Reservoir in Hoshangabad region

(Source-STR office file)

From cultural anthropological point of view, history of Pachmarhi region is a history of blatant destruction of biodiversity as well as its ecosystem and unprecedented violation of human rights. As per local legend, even 200 years ago, this area was covered with dense luxuriant forest, rich in flora and fauna and contained numerous human settlements in the form of tribal villages. There had been a good harmony among man, wild life and forest. Forest of this area has another uniqueness. Usually Sal (shorea robusta) and teak plants grow in different ecologies. This area happens to be the boundary between these ecologies where one can find both Sal and Teak forests with mixed forest in between (Singh, 2003:106).

There are many caves in this area; a few of them exhibit cave the paintings. It is not known whether any scientific study have been made on them. But some scholars in their writing have stated that the paintings may be 10,000 years old (Pathak, 2003:89-95; Singh, 2003:103-121). These paintings are of different types of wild animals like bison, pig, deer, elephant, tiger, etc. Hunting scenes are also common where the hunters are shown in small groups with sticks, bows and arrows and spears. Besides these, human figures are also shown in different activities such as dancing, mother nursing her baby, honey collection, fighting scenes, and so on. Most of these paintings are made up of natural colours. In some cases, abstract symbols have also been noted. The Mawasi and Gond mainly inhabit this region. The Mawasi tribals living in this region make memorial board (patia) made of wood. These boards are kept under a sacred tree in memory of the dead relatives. These boards contain many figures like human beings, horse rider, sun, moon, tree, hut, and so on. Styles of the engravings of the memorial boards and the figures, which are observed inside the caves, have striking similarities. It may be said that perhaps forefathers of the present day tribals living in Pachmarhi region might have made these cave paintings. Cave paintings can be found in various conditions, some are in good condition, some are faded, some are disfigured by human vandalism and many of such paintings are covered with thick layers of mineral depositions and moss which require immediate attention.

As per local legend, the Pandava Princess of Mahabharata fame had visited this region during their exile. There is a cave in Pachmarhi town in their name. In 1860, representatives of the British Empire started visiting this place. In those days, the British rulers were busy strengthening their position in the entire Madhya Pradesh region, which then was known as Central Province. Captain J. Forsyth first came to the valley of Satpura Hills. He liked the place for its climatic condition, ecology, topography, and so on. They constructed a lodge type house at Pachmarhi and named it as Bison Lodge and now it is a Museum. Gradually a small hill town grew up along with a military cantonment, a small market and a church. Later on, many bungalow type houses were constructed for the residential purposes of the British administrators. This plateau was the summer capital of the Central Province during British period. Remains of the past era are still found in the Pachmarhi. In 1862, the British rulers established their Forest Department in India and observing the dense forest in the region, they decided to extract timber for commercial purposes. They first demarcated Bori Forest and declared it as a Reserve Forest. The Bori Reserve Forest, which is a part of present PBR, happens to be the first Reserve Forest in British India (Ref: Gazetteer of India, M.P. Hoshangabad, Bhopal 1997, 205-206). Gradually the other forests were also demarcated and declared as Reserve Forest, Protected Forest and Village Forest. Many regulations were imposed on the indigenous tribal societies who were living in those parts. Within a short period of time they lost their ownership on surrounding forests. Their economy of hunting and gathering inside those Reserve Forests were totally banned. This is a gross violation of human rights. Secondly, the British people started felling of trees for extraction of timber for commercial purposes. This process continued till 1947. From 1889 to 1947, a great destruction of the eco-system prevailing over there was complete.

A summary of the historical background of the forest (Gazetteer of India, Madhya Pradesh, Hoshangabad, 1997:205-206 and EPCO, 1996:15-18) of this region is given below.

The old Bori forest located in the east of the Sonbhadra River was the first forest in India to be declared as a Reserve Forest. This area was confiscated in 1859 from Thakur Bhabut Singh (a Gond Chieftain) for his disloyalty during Sepoy Mutiny. As stated earlier this area was brought under the management of the newly organized Forest Department in 1862. During that period, the local aboriginal tribals used to practice shifting cultivation inside the forests, locally known as 'dhaya'.

Special measures for protection were taken in 1864 and in the following year the forests were notified as reserved under the Act of 1865. Four years later, a resident forest officer was placed in charge with headquarter at Harapala. Boundaries were demarcated and at Harapala a small teak plantation was established. In 1878, in order that a larger teak producing area might be included in the reserve, the new Bori forest, stretching from Sonbhadra to Kankri was added to the Bori range. Until 1889, practically no felling was attempted. Since then, annual cuttings of mature teak and tinsa (ongenia dalbergioides) have been made in defined areas according to a systematic working plan. The forests of the other four ranges (i.e. Kamti range, Pachmarhi range, Bhoura range and Matkuli range) were first brought under government supervision in 1862 when they were notified under the wasteland rules. The cutting of certain reserved trees namely teak, saj (boswellia serrata), shisham (dalbergia latifolia) and bija (pterocarpus marsupium) were prohibited; but subject to the provision of free access were granted on payment of an annual commutation fee and large areas were leased out in block. The need for closer control was soon apparent and in about 1868, some more forests were notified as reserved and brought under Forest Department. The remainder designated as unreserved forests were managed by the District Authority until 1878. Later on, those too were handed over to the Forest Department. Commutation, except for the aboriginal hill tribes, was abolished and the license system introduced. Forest produce was only permitted to be removed from defined areas by fixed routes and each removal had to be paid for separately.

In each biosphere reserve, there are different zones which are carefully demarcated for the proper conservation of its natural flora and fauna, such as : core zone, buffer zone and contact zone. Core zone happens to be the 'gene pool' area and requires being free from any type of human manipulation. Human settlements are allowed in the contact zones having buffer zones between them. PBR has the following zones.

(a) Core Zone: Present core zone of PBR is about 524.37 sq. km. (i.e. the entire Satpura National Park area)

(b) Buffer Zone: An area of 4462.93 sq. km. around the PBR core area forms the buffer zone.

PBR has no contact zone and the other specialty is that, thousands of years before present (as evident from the published literatures), the entire area was inhabited by various ethnic groups among whom the tribal groups (namely, the Mawasi, Gond and so on) were the major settlers. Before declaration of this particular area as biosphere reserve, the concerned authorities should have considered this fact with utmost care; which could have saved the local population from a lot of humanistic problems. In tribal villages, village land, forest, water sources etc., are common properties (CP) of the village. Villages mainly have usufructuary right only. During the declaration of Pachmarhi Sanctuary and Satpura National Park in 1977 and 1981 respectively, all the revenue villages situated in and around Satpura National Park had to lose their common properties. In some cases it was more than 98% thsat too without any compensation.

This type of violation of human rights in post-independence era will be elaborated further in the later chapters.

Published literature on the PBR by its nodal agency states the following physical features of this region.

'The altitude varies from 320 to 1352 m. Satpura hill ranges run east to west. The general configuration of the area is hilly undulating terrain and at places precipitous having deep gorges around Mahadev Hills in Pachmarhi plateau. The Satpura hill ranges run east to west and the Pachmarhi Plateau is practically in the centre of the area having an elevation of around 1050 m. The Pachmarhi hills have steep slopes in the north and in the south it is straight exposing the bare rocks. Some of the conspicuous hilly areas are Jambudweep, Dhoopgarh, Handikho, Mahadev, Chauragarh and Bee Fall. Amongst these, the Dhoopgarh is the highest point (1352 msl) in the State. The area exhibits variety of geological rock and soil formations. The Archean Phyllites and Schist's, the Jurassic, Permian Gondwana sandstone formations of Bagra, Denwa, Pachmarhi series, Cretaceous, Eocene, Deccan trap sills and Dykes and recent Alluvium are found in the area. PBR has typical monsoonic climate with three distinct seasons: summer rainy and winter. The variation in climate occurs from place to place. The Pachmarhi plateau is cool in summer and has heavy rainfall in rainy season, whereas low lands in Narmada basin are uncomfortably hot in summer with less rainfall. The mean daily maximum temperature ranges from 26^0 to 42^0 C and minimum 9.7^0 to 25^0 C depending upon the season. The land use of the area based on Landsat Imagery of April 1990 and November-January 1991-1992 comprises forest 65.2%, agriculture 27.7%, water bodies 4.2%, wasteland 2.5% and built-up land 0.5%. Of the total forest cover, the closed forest constitutes 85.3%, open forest 8.2%, degraded forest 4.2% and forest blank 2.3%. The Tawa reservoir is the major constituent of the water bodies. The close forest constitutes 85.3% of the total forest cover while the open forest cover is 8.2%. The degraded forest constitutes 4.2% of the total forest area, while the forest blank is 2.3%. Of the total wasteland, the gullied and ravenous lands constitute 1.4% only and the rest 98.6% comprises land with/without scrub. Of the total areas covered under water bodies, 30.2% fall under river/streams and the rest 69.82% falls under reservoir and tanks' (DOCI-2001, BRIS 1:1,10-11).

Promulgation of Indian Forest Act was done by the British in 1865. This was again superseded in 1927. Both the aforesaid Forest Acts were not in favour of the tribal groups whose native villages are located in and around the reserve forests. With a few amendments, the Indian Forest Act of 1927 is still being followed in the cases of all types of forested areas.

Social scientists emphatically state that the tribal groups have rich cultural traditions, which requires to be protected at any cost. They should have control over their natural resources and any type of profit generated from there should go primarily to them. Their ownership over their natural resources should be protected and so on. In fact in the 'Tribal Panchsheel Policy' (Nehru, 1960) also similar types of policy statements have been made. But in reality, what happens in a tribal area is quite different. 'The invasion of Adivasi territories which for the most part commenced during the colonial period, intensified in the post-colonial period. Most of the Adivasi territories were claimed by the states. Over 10 million Adivasis have been displaced to make way for development projects such as dams, mining, industries, roads, protected areas, etc.' (Bijoy; 2003,5). Condition of PBR is no different at all. The Satpura National Park, which is the 'Core Zone' of the PBR, contains nine villages predominantly inhabited by the tribal communities. Perhaps, demarcation of 'core area' was done without considering this fact, in haste.

(e) Methodology

Based on the information collected from Bhopal, it appeared that the situation in PBR core area is highly complex. It compelled us to formulate our objectives in such a manner so that it may not become objectionable to the Madhya Pradesh Forest Department, which actually looks after the area in collaboration with the Satpura Tiger Project authorities. Our field investigation had to face a lot of difficulties due to some terms and condition, which were imposed upon us through the Madhya Pradesh Forest Department's permission letter, which allowed us to enter into the Satpura National Park (SNP). Out of nine clauses mentioned therein, the following had directly affected our field investigation.

(a) Researchers are allowed to undertake investigations on 'Ethnological Aspects' only. At present rehabilitation work of villagers from Pachmarhi and Bori sanctuaries is in progress. They should never interfere in that work, nor they are allowed to discuss anything with the concerned villagers or any Non Governmental Organization (NGO) in this respect (Clause-3).

(b) Research investigation may be carried out between sunrise to sunset. Without permission of the Field Director, nobody is allowed to enter or remain inside the reserved forest at night (Clause-5).

(c) Researchers are not permitted to write any comment in their reports, which may criticize forest department's activities (Clause-6).

(d) For necessary approval, Anthropological Survey of India should submit two copies of the report to this office (Clause-7).

The aforesaid restrictions were unexpected for a Central Government Research Organization like Anthropological Survey of India and dampened our spirits to an extent but at the same time confirmed our collective doubts that, perhaps, things are not in proper shape inside the PBR core area. We also had no option to return back in protest without doing our job. In that case, we would have to give explanation to our competent authority and face his displeasure. We held an emergency meeting among us and decided to go ahead. Wherever we went for field investigation, we noticed forest department employee's following our team on some pretext or other. This made us more careful about our action, speech and behaviour.

We had to stay at a hotel in Pachmarhi town located on the border of the PBR core area. Roads inside PBR core area, mostly unmetalled, are in many cases so difficult that only 4-wheel drive petrol jeeps can pass over them and that too after crossing several hill streams and rivers. Our team hired one of such vehicle at Pachmarhi. That too, one of the Deputy Forest Ranger posted at Pachmarhi requested us to select a particular vehicle. Again we had no option, as we were unaware about how much we would have to depend on the forest department during our fieldwork. We took some resolutions among ourselves that, while moving by that vehicle or in the presence of its driver, we must not discuss any topic, which may go against the policies of the forest department.

In core villages of PBR, field data were collected by household census, observations, collection of case histories, interviews and through group discussions. Genealogy of all the families was collected to find out their migration pattern, marriage distance, and so on. To supplement our findings, a good number topic wise still photographs have also been taken.

1

Pachmarhi Biosphere Reserve : Core Area Study

S. Sengupta, D.N. Pandey and A. Roy

Cultural Background of the Indigenous People Living in the PBR Core Area

The entire 'core area' of PBR, from time immemorial, was inhabited by indigenous tribal communities. It contains forested high plateau of the Satpura range where the hills run from east to west with intermittent valleys and thus forms a very beautiful landscape. "The *Ramayana* and *Mahabharata* speak of the entire tract ...... as a land of wilderness inhabited by demons, while religious hermits of Aryan races dwelt in hermitages in their midst ...... In the fourteenth and fifteenth centuries, many Rajput clans, who by Akbar's time probably had succeeded in reclaiming the Narmada valley for agriculture, infiltrated this vast land of the aborigines. The Gonds retired to the higher plateaus or slopes, to continue their hunting, and their own method of "Dhaya" or shifting cultivation. In course of time, even the larger plateaus were invaded by the aggressive northerners. The aborigines remained only in charge of those lands, which was too arduous to till by the northerner's methods (EPCO-2001;1,11)."

The British rulers took interest in this region in the nineteenth century. In the beginning they were interested in the fertile plains and did not take any interest in the forest dwelling tribes and their habitat. After the mutiny in 1857, they wanted to strengthen their position in this region, which resulted in the creation of the Central Provinces under the supervision of a Chief Commissioner. In 1862, Capt. J. Forsyth first entered into Pachmarhi plateau. In 1870, a cantonment was established there. Due to its cold climate during summer, gradually Pachmarhi became the summer capital of the Central Provinces during British rule. They fully explored all the areas in and around Pachmarhi and identified all the places of tourist interest and made them accessible to the tourist or visitors. A lot of these places are now inside the PBR core area.

There are nine villages in the PBR core area. Out of them eight are 'revenue villages' and the rest one is a 'forest village'. * They are mostly inhabited by two tribal communities, viz. (i) the Gond, and (ii) the Mawasi. Besides, a few families of Yadav, Thakur (Rajput) and Muslims live in a couple of villages. They are outsiders who settled there at a later period. So in this chapter, we would describe background history of the Gond and the Mawasi only.

Gond

The word ' Gond' perhaps has a non-tribal origin. Some people say this term might have its origin from the words 'Gund' or 'Kund'. 'Kond' is a Telegu word which means mountain (Agarwal-1999,300). In the opinion of scholars, they belong to the Dravidian stock and are the second largest tribe in India. In the state list of Madhya Pradesh, Andhra Pradesh, Maharashtra, Bihar, Gujarat, Karnataka, Orissa and West Bengal, they have been included as Scheduled Tribe. In Madhya Pradesh, they are the largest tribal community with higher concentration in the Mandla district. The Gonds of Madhya Pradesh speak 'Gondi' dialect as their mother tongue. They also speak Chhatisgarhi dialect and Hindi to communicate with others. In the opinion of the Anthropologist, 'Gond' is a generic term given to a large group of tribal people living in and around Madhya Pradesh and are divided into many endogamous sub-groups. There are many pubiished literature on them.

---

* Differences between the 'revenue village' and the 'forest village' will be discussed in detail in the next chapter.

Gond settlements in PBR core area are usually small and numbers vary from 20 to 40 odd families. Settlements are uniethnic in nature, except in Bariaam village where a few Rajput and Yadav families are observed. It was told that the traditional village panchayat had approved their stay over there and had given some village land to construct their houses. Gonds of this area refer to themselves as 'Gond Thakur' whereas the people of neighbouring area (such as in Pachmarhi town) simply refer to them as 'Gond' only. In the past, the Gonds had some ranking system, i.e., vertical stratification in their society. There were nobles among them in the form of king, chief, priest, etc., where eldest son inherited these offices after their demise. The rest of the population, divided into many exogamous clans had equal status. Now with the decline of kingship and changed political scenario around them, the aforesaid stratification based on hereditary ranks have been reduced to irrelevance. However, their age old exogamous social units, namely family (parivar), lineage (vansh), clan (gotra) etc., have retained their importance (Banerjee-1998,287-290).

A Gond village has a main village street with houses arranged serially on either side and facing towards the streets. All the homesteads have fencing to demarcate their boundaries with an opening as entrance. Houses are usually mud built and thatched with straw or long type of dried grass available in the jungle. Some houses are also thatched with earthen tiles. In some houses, sidewalls are made of stone blocks cemented by clay. Each household has a large living room (bhitari) with/without window. In one corner they cook, store their grains in mud plastered bamboo baskets and in the other corner there would be a raised platform for their family/clan deities. Each house has a thatched spacious patio (usari). Patios may be opened or may be closed on two sides. Adjacent to the living room and patio, there would be a cattle shed. There would also be a raised bamboo platform where they keep their water pots. On the backside there may be a small kitchen garden or a cage to keep poultry birds.

In each village, there would be some shrines for the village deities, viz. Hardul Dada, Gowal Baba, Bajrangbali or Hanuman Dada, Kherapati Bai, Siddha Baba etc., who protect the villages and their domestic cattle or crops from any kind of harm or calamities. Their shrines are constructed in the open under a tree, where a boulder smeared with red ochre (sindur) symbolically represents the deity. In some cases, instead of boulder, iron tridents smeared with red ochre have been found. These are the important cultural landmarks in a Gond village.

Among the Gonds of PBR, descent is recognized by father's line only. Clan affiliation of the girl's change after their marriages. Gond families are nuclear in type. Father acts as the head of family. In case of the father's death, mother acts as family head if the children are young. Otherwise, the adult eldest son takes over the charge. There is no great gender discrimination among them. Sons and daughters are considered almost as equal. Sons have a slight edge over the daughters as they perform post funeral rituals.

In the PBR core area, the Gonds consider themselves as superior to the Mawasis. There is no connubial and commensal relation between them. Hypergamy is allowed whereas hypogamy is very strictly dealt with. If a clean caste Hindu person, eg. Rajput, Kurmi, Katia, etc., gets ex-communicated from their caste panchayat for marrying a Gond girl, he will be accepted in the girl's village and the village panchayat would help that person by offering land so that he may be able to construct a house and live there without any discrimination. If a Gond boy marries a clean caste girl, he would not face any difficulty in his native village. But if a Gond boy marries a Mawasi girl, he will be ex-communicated. He will either settle in the girl's village or would settle elsewhere in any urban area where his marriage will not be questioned by anybody. The Gond boy can return back to his village if his wife dies or he divorces her. In both the cases, he has to perform purificatory ritual as per their social custom and has to give a communal feast. In case of his death outside as also his wife, his children may come to his native place and beg for their settlement. In such cases, the village panchayat usually do not object and the children enjoy equal social status in the village society.

In PBR core area, Gond clans are exogamous, of equal status and are characterized by the number of deities they worship. They never utter the names of their family or clan deities, not to speak to an outsider even not to another fellow Gond. They only mention their numbers. Numbers of clan deities vary from one to eight. For marriage negotiation, one party asks the other party about the number of clan deities. If they happen to worship same number of deities, then marriages will not take place. Clans which worship odd number of deities (i.e., 1,3,5 and 7) are collectively referred to as Une groups. The worshippers of even number clan deities (i.e., 2,4,6 and 8) are referred to as Purhe groups. Marriages among the Une clan members are not allowed as they firmly believe that they have consanguine relation (albeit cannot be traced through genealogy) among themselves, i.e., *bhaibandhi*. Similarly the Purhe clans also do not intermarry due to the same social restrictions. Marriage alliances are only allowed between the Une and Purhe clans. During field investigation in the PBR core area, the following clans were found among the Gonds.

1. Une Clans: (i) Arpanche, (ii) Bariba, (iii) Dhurbe, (iv) Erpachi, (v) Imne, (vi) Kakoria, (vii) Marskoley, (viii) Naure, (ix) Padram, (x) Sarbeyan, (xi) Saratia, (xii) Siyarsaran, (xiii) Vallabey.
2. Purhe Clans: (i) Barkare, (ii) Barkey, (iii) Batti, (iv) Eke, (v) Kumre, (vi) Porta, (vii) Tekam, (viii) Ukey.

Gonds are monogamous though polygamy is not prohibited in their society. Cross cousin marriage (MBD) is also practiced. Acquiring mate through negotiation is the usual practice. The guardian of the boy along with a few relatives and village elders visit girl's house with the proposal. The girl's guardian invites some village elders and the marriage proposal is discussed there and the marriage date is fixed. Gond marriages do not take place in the month of Pus (December-January), Fag (February-March) and Chet (March-April). Ujli paksh (i.e. full moon) period is considered to be auspicious period for arranging marriages. Marriage takes place in a specially constructed arena on the courtyard of the bride's house. Consummation of marriage takes place in the bridegroom's house.

Marriage by mutual consent and running away from the village is prevalent among the Gond. Locally it is referred to as 'love marriage'. After one year or so, the couple may return to the village. They have to beg apology in front of the village council. If the clans of the couple match with their social norms, they are included in the village society. Bridegroom has to arrange a communal feast for the same. If their clans do not match, then they have two options. First one is dissolution of the marriage and the concerned persons go back to their respective families after paying penalty. The other option is, the couple again run away and settle in such an area where nobody cares for their identity.

Among the Gonds, a girl child is never discriminated. They are valued possession instead for the type of service they render to their family. As a result, bride price is high among them and has to be paid both in cash and kind. If a youth does not have sufficient money to pay bride price, even then he may approach a girl's father to marry his daughter. If the girl's father finds the youth suitable to be his son-in-law then he invites the village elders. In their presence they make an agreement that the youth would work for the girl's father as farm labour for 2-3 years without any remuneration. He will be given meals twice daily and a place to sleep at night. During this period the youth, however, is not allowed to meet or talk to his would-be wife. If he can fulfill his agreement, the girl's father arranges for their marriage. The youth then gets two options. He may return back to his native village or settle in the wife's village. This is locally termed as *lamjane*.

If a man has an only daughter, then he may wish to keep his son-in-law in his own house. He may approach a suitable youth for his daughter. If the youth.agrees then marriage is solemnized in the girl's house. This is known as *palak-beta sadi* (marriage by adoption of son-in-law or marriage by staying in father-in-law's house).

Junior levirate is also found among them. A Gond youth can marry the widow of his elder brother if she is in marriageable age and willing to that alliance. However, there is no social compulsion. But a man is not allowed to marry his younger brother's widow. Remarriage of divorced woman and widows are permissible. This type of marriage is locally termed as *ghar-baitha sadi*.

Divorce is permitted. Either of the couple can seek divorce. It is granted in a meeting attended by the village elders of both sides. If the wife is found guilty of adultery or any type of gross misconduct, then the bride's parents have to return the bride price. Otherwise the bride price is forfeited.

In a family both the couple work equally, either in their agricultural field or outside as a day labourer and contribute accordingly to their family income. Important decisions are taken jointly, though only the husband can take part in the village council meeting. Regarding size of the family at present, it was reported that, they consult each other and take joint decisions. Conception is considered as a natural phenomenon and the pregnant ladies do not face any restriction in food or movement. Only in the advanced stage, they are not allowed to participate directly in village rituals. Gond children are born at home. They do not go to hospital for periodical check-ups and tetanus toxoid injections. Older female relatives help in the delivery of the child. Umbilical chord is cut by a sharp knife and is tied by a piece of thread. Placenta is buried inside the labour room. The old lady, who acts as a midwife, digs a hole in the floor and disposes it there. Depending on the health condition of the parturient mother new born baby gets its first breast feed after one or two hours. They feed colostrums. On the fifth day after childbirth, all the female folk of village assemble there for merry making. They sing song and bless the child. They are offered gur (molasses) on that occasion. Mothers are provided with normal food. Often her diet is supplemented with kutki (a kind of rice), tur dal (a type of pulse), safed musli (a kind of root with medicinal value), dalia (coarsely grounded wheat grain), ghee (clarified butter) and kismis (raisin) for an early recovery. The parturient mother is kept secluded with the newborn for six weeks. She is not allowed to do any household work. After this period, she takes a purificatory bath, worships her household deities and resumes her domestic responsibilities as before.

Bringing up of children within the family is primarily the job of the mother. Other family members, specially the elder children, help her in this respect. When the mother goes out in the jungle for collecting fuel wood or any non-timber forest produces (NTFP) or to work as daily wage labour, these boys and girls look after their younger siblings. As they grow older, they get further training from their elders. Usually the mother and the elder female members train the girls about their social and economic roles and responsibilities. On the other hand, the boys are trained by their fathers and the other elder male members in the family. Usually at the age of 16 years, they come out as an earner and contribute to the family fund. They usually dispose the dead body by burying. Occasion of death in anybody's house becomes a village affair. All the kins come forward and offer their help and co-operation. They carry the dead body to the morghat (burial ground). It is generally located near a stream. While taking the dead body to burial ground they sprinkle *lajmurmuri*, which contains puffed rice and small coins on the entire path. At the graveyard, they give a bath and thereafter place Panchmewa (assorted five types of dry fruits, namely copra, raisin, cashew, almond and groundnut) in the mouth of the deceased as last food. While these are being done, the other people dig a grave. The shrouded body is then lowered into the grave and its head remains in the direction of the stream. Everybody puts a fistful of soil in the grave. Then they fill up the grave with soil. They also cover the grave with big stone boulders so that wild animals may not be able to dig into it.

Apart from burial, they may even dispose the dead body through cremation, locally termed as Dag. Perhaps they have adopted this practice from the neighbouring Hindu people, though its frequency is less in comparison. In case of cremation, they collect rakh (ash) from the cremation spot after three days. It is kept outside in a flat dish for a whole night. Next morning they check that ash. If it contains any impression or footmark of any bird or animal, it is believed that the departed soul will have rebirth in that form.

Pollution period due to occurrence of death lasts for three days. It is known as *teeja*. All the villagers and the relatives of the deceased visit the graveyard or cremation ground (as the case may be). A special dish consisting of *kanki* rice, vegetable curry, fruits and *mohua* liquor is offered there in the name of the deceased person. It is believed that, after being content the soul of the deceased starts its journey to the other world. After performing this rite, everybody returns back to attend the communal feast arranged by the family members of the deceased.

The Gonds of the PBR core area believe in many gods and goddesses and a few of them also belong to the Hindu pantheon.

For example, they consider 'Shankar Bhagawan' as their prime god and 'Durga Devi' as prime goddess. Besides, they worship:

(a)	Bari Mata or Bari Devi	-	Forest goddess, to protect them inside forest
(b)	Chotti Mata or Chhoti Devi	-	Forest goddess, to protect them inside forest
(c)	Kherapati Bai	-	Forest goddess, to protect them and their agricultural fields from the wild animals
(d)	Siddha Baba	-	Village god, to protect their drinking water sources
(e)	Gowal Baba	-	Village god, to protect their domestic cattle from wild animals
(f)	Saj Ka Ped	-	Tree god, a valuable auspicious tree which has properties to protect them in any sort of problem'
(g)	Hanuman Dada or Bajrangbali	-	Monkey god, protects them and their abode from any type of distress
(h)	Seyane Burha	-	Ancestral god, to protect their families from any mishaps
(i)	Bagh Deo	-	Tiger god
(j)	Nag Deo	-	Snake god

The Gond also believe in malevolent or evil spirits like, the 'churel', 'dakri' or 'dain', etc. These spirits are believed to reside usually upon the trees in the jungle. They may cause harm to anybody who unknowingly passes under their abode and make them sick. To get rid of such afflicted sickness they usually go to a shaman (parihar).

Following the neighbouring Hindu castes, they also observe Diwali, Maha Shivratri, Nag Panchami and Navratri festivals and attend fairs organized at the nearby urban centres.

Mawasi

In terms of population number the Mawasi are the largest scheduled tribe population in the PBR core area. They live in Badkachar, Binora, Kajri, Ghoranar, Pagara Dhana and Rorighat villages. Besides, a couple of Mawasi families live in the multi-ethnic Bariaam village. According to the published literature, the Mawasi happen to be the largest sub-group of the Korku tribes of Madhya Pradesh (Russel and Hiralal-1916,551; Tiwari and Sharma-1994,74-75). They are also referred to as 'Mowasi'. It means 'the inhabitant of Mowas', a country which in the opinion of the Marathi people is situated in the western part of the Satpura Hills. To them, it meant a 'troubled country' residents of which occasionally used to come down from their hilly areas to the rich river valley areas for plundering. This happens to be a similar situation, which prevailed, in the pre-British period between the Nagas of the Naga Hills and the Assamese population living in the Brahmaputra plains.

When we asked the Mawasi whether they had any link with the Korku, they declined any such linkages. In their opinion, they were a separate group. Since our time was limited for the completion of the present study, we did not spend much time in finding out their history of origin.

Mawasi are an endogamous tribal community divided into many exogamous 'gotras' or clans. In the PBR core area, we found the following exogamous clans, viz, (i) Basaria, (ii) Basunda, (iii) Bhopa or Baithak, (iv) Bhusunda, (v) Darsama, (vi) Dhiku, (vii) Loba, (viii) Kajla, (ix) Kalam or Bhabai, (x) Kasda, (xi) Sakam, (xii) Silu, (xiii) Rajkar, (xiv) Takhan and (xv) Valem. However, it is felt that, the above state gotra list is not exhaustive. Perhaps in the Mawasi villages located in the PBR buffer zone areas many more such gotras would be found.

Mawasi society is patrilineal where descent is recognized through father's line. Alike the Gonds, as mentioned before, the Mawasis also of this area refer to themselves as Mawasi Thakur. However, the other neighbouring ethnic groups describe them as Mawasi only.

Properties such as agricultural fields are inherited by the sons only. Families are mostly nuclear type. Married sons usually stay separately. Aged parents often stay with the youngest son but this is not a standard, i.e., customary practice. Types of marriages appear to be the same as mentioned in the case of Gonds before and their local terms are also the same. For example, the Mawasi also use the term *lamjane* for marriages by service. However, their rituals are much more elaborate and in some cases differs from the Gonds.

The local term for pregnancy is *amal*. It is a natural phenomenon to them and they do not attach any special importance to it. Pregnant ladies do normal household duties, which includes going to the forest for collection of fuel wood and other NTFPs, upto 28 weeks. They take normal food. They only do not take carrot, radish and guava as these are considered to possess power to inflict cold, which may harm the foetus. Delivery of child is called *jachki*. Children are born in the house. A midwife or *dai* helps in childbirth. An old lady or female relative, having enough experiences of childbirth, usually acts as midwife. They never go to any health centre or hospital for this purpose. Our investigation also reveals that, to avail such facility they have to travel a long distance, which is inconvenient as well as expensive for them.

The midwife severs umbilical chord (nara) of the newborn with an iron knife. It is then tied with a thread. Powder of earthen tiles (khapra) used for thatching roof, is applied on the umbilical stump. The midwife digs a hole in the floor of the delivery room to dispose off the placenta. She then gives a warm water bath to both the mother and the baby. Other female relatives also help her in this process. After that, the midwife washes her hands and feet, lights incense sticks and offers a coconut to the household deities as an act of thanksgiving for a trouble free delivery and also seeking their blessings for the newborn child and its mother. This process is known as *sudak*.

In the past, newborn babies were not given colostrums. Following the advice of the village level health workers who visit their villages' occasionally, they now give colostrums to the baby. Babies are fed on demand. Breast-feeding of a baby continues as long as its mother can provide, i.e., till the onset of next pregnancy. If any mother suffers from insufficient lactation, her child is given diluted goat's milk. They cut garlic cloves into small pieces and a tablespoon full of such pieces is deep fried in mustard oil until they become brown in colour. The oil is then cooled and the garlic pieces removed. This oil is then used for massaging the newborn child. In their opinion, this massage prevents the attack of cough and cold and makes the bones stronger. For a week, the parturient mother is given special diet so that she can recuperate quickly. Powdered *ajowain* (thyme) is added to dough to prepare roti (handmade bread) for her consumption. Besides, she is also given special preparations of suji (a kind of coarse wheat flour), dalia (coarsely broken wheat grain), ghee (clarified butter), kaju (cashew) and kismis (raisin). Some other local herbs are also sometimes given depending upon mother's health condition names of which they did not tell in spite of our queries.

The parturient mother has to remain confined in the labour room for one and a half months. After that *chokpai* (purificatory bath) is given to her so that she may be able to return to her normal life. Before the purificatory bath, mother is not allowed to come out and do cooking or any other domestic work. She is also not allowed to worship the household deities. This occasion is known as *badhai*. During this celebration, relatives such as maternal grand parents, maternal uncles, and father's sisters along with other lineage members are invited. The guests bring presents and sweetmeat for the occasion and bless the newborn child. They unanimously select a name for the baby. They are entertained with kanki rice, roti (handmade bread of wheat flour), laddu (a kind of sweetmeat), patasa (sugar wafer), gur (molasses), bundi (a kind of fried substance dipped in sugar syrup and of course, with lots of mohua liquor. The midwife is also invited in this celebration. She is given *sidha* in a basket which contains rice, a coconut, pulses, salt, dried chilly, pieces of turmeric, a sari (loin cloth), a blouse and Rs. 5/- or Rs. 10/- in cash, as per their tradition. In this ceremony, they also invite five young girls from the village who have not attained puberty. They are fed ceremonially and the baby's father pays them Rs. 5/- each.

In Mawasi family, there is no discrimination between boys and girls. They are treated equally and their socialization process is similar to that of the Gonds of this area, as has been mentioned before.

Negotiation is the most preferred for acquiring a mate. One person who may belong to either of the party's acts as a go-between. During negotiation, everything is discussed between the two sides in the presence of their respective village council members (sheyane), usually elderly people, and then the would-be bridegroom and bride see each other. If they give their consent then the elders finalize the negotiation and present a note of Rs. 50/- or Rs. 100/- to the bride. This is called *pakka* or *hathlia* (i.e., reservation of the bride). After this, the bride's guardians do not have any right to settle her marriage elsewhere.

The next phase is *sagai* or betrothal ceremony. It is performed as early as possible. It takes place at the bride's house. The bridegroom's side is represented by his mother's brother, mother's sister's husband, sister's husband and a few elderly persons from his village (usually lineage member). In sagai, the bridegroom's side is called *ganga*. The bride's side is represented by her brother's wife, mother's sister, father's sister along with their spouses. They are called *sabasan*. Main function of the sagai is to present a silver engagement ring to the bride. Any of the female members of the *sabasan* accepts the ring from the ganga group and insert the ring to the finger of the bride. A basket containing coloured sari (loin cloth), polka (blouse), kangi (comb), mang/sindur (vermilion), a bottle of hair oil and a few pieces of silver ornaments, etc., is also presented to the bride. Sagai is completed with a feast along with country liquor. After this, actual marriage may take place even after one year as the ceremony is expensive and both the sides have to make some preparations for that.

After the finalization of the marriage date, *parcha* (formal invitation) is sent to all relatives and to all families in the village. Parcha consists of a betel leaf containing a vermilion dot, areca nut, a few grains of rice and a small piece of turmeric. About two weeks before the marriage date, special protections are given to both the bridegroom and the bride to protect them from any type of evil spirits. To ward off harmful effects of evil eyes or evil spirits, the bridegroom carries a sickle with him in which a piece of red cloth remains tied to the blade. Similarly the bride carries areca nut cutter wrapped up in a piece of red cloth.

During this period, *marowa* (special arena) is prepared in bride's house. A corner of the courtyard is selected for this. Surface is made smooth and even with several layers of cow dung and mud paste. Surrounding a central wooden pillar of a Sal tree, nine other pillars are erected. The upper ends of these pillars are joined with each other with the help of split bamboo. These bamboo splits form the infrastructure of the roof of Marowa. It is then thatched with green leaves of mango and many other trees, which are locally available.

Barat (marriage party) reaches the bride's house one day before the ceremony. Previously from bridegroom's village, the marriage party had to walk down the whole distance to the bride's village or in some places by bullock cart. At present, due to improved transportation system, they often use hired vehicle. In the past, barat used to reach the bride's village well in advance and used to bear all the expenses of their food and drinks themselves till the previous day of the marriage ceremony. At present, they reach one day before the marriage ceremony and all their expenses on food and drinks are borne by the bride's family. The youngest member of the bridegroom's party is called *soasa*.

Marriage dress has special colours. Bridegroom wears yellow long shirt (kurta) and a folded piece of white cloth (1.25 m in length) which is kept hanging from his shoulder while the bridal dress consists of red or pink sari (loin cloth) and blouse (polka). Both of them wear a special type of headgear made of *chhinda jhar* (a particular type of bush).

Properly dressed bridegroom and bride are then brought to the decorated *marowa* and they take their seats on wooden planks side by side. The elderly persons assembled there start blessing the couple serially. They mark their forehead with some coloured material and offer gifts either in cash or kind or both.

After this, actual marriage ceremony starts. At first, the *parihar* (village priest-cum-shaman-cum-medicine man) worships *kherapati bai*, a local deity. He lights an oil lamp, incense sticks, and with the recitation of incantations offers six different types of fruits to the deity. After invocation of kherapati bai, the marriage ritual begins. Both the bridegroom and the bride circle around the central pillar of the marowa in clockwise fashion and tie dhaga (coloured thread) to the pillar. They do it five times amidst the *vamra* (i.e. the recitation of incantations) of the parihar (priest). By that way the first phase of marriage is completed. The parihar receives a dhoti (loin cloth for male), *patasa* (sugar wafer), gur (molasses) and some cash as his remuneration. This custom is known as *pitgiri*. At the time of marriage ritual, the bride's father ceremonially offers five brass utensils to the couple as gift. They are khagre (bowl), thali (plate), katori (small bowls), lota (water pot) and chamach (serving spoon).

The next day after marriage feast, the marriage party returns to the bridegroom's village with the bride. Here the second part of the marriage is solemnized. A *marowa* is constructed in the courtyard of the bridegroom's house. Here again amidst *vamra* they take two rounds around the central pillar in the same manner as stated above. Altogether seven circumambulations of the central pillar are needed to complete a Mawasi marriage. On the next morning, the bride goes back to her parent's house along with her relatives. After a couple of weeks, the bridegroom's relatives go to the bride's village to take her back. Afterwards, the suhagrat (i.e. consummation of marriage) is celebrated. In the traditional Mawasi marriage, there is no custom of exchanging garlands between the bride and the bridegroom, which is an essential part of Hindu marriage in the neighbouring areas. In very few cases, the Mawasis at present go to Hindu temples to get married over there, mainly to cut down marriage expenses. As in these cases, they have to exchange garland so they are termed as *jaimala shadi*.

Alike the Gond, the Mawasi also practice lamjane (i.e. marriage by service) type of marriage with similar kind of terms and conditions. During survey, 19 cases of *lamjane* marriages have been observed (Pagara Dhana-4, Ghoranar-1, Badkachar-6, Jahru-4, Rorighat-2 and Binora-2).

Beside, *palak beta shadi* (Adoption of son-in-law in the family), *pat ruti shadi* (marriage by elopement) *ghar baitha* (remarriage of widow or divorced females) and junior levirate are found among the Mawasi. Customs are also similar to that of the Gonds, as have been mentioned earlier. During survey, four cases of *ghar baitha shadi* have been found (Kajri-2, Rorighat-1, Binora-1). Single case of junior levirate was found at Binora village.

The Mawasi are usually monogamous but having more than one wife (i.e. polygamy) is not prohibited among them. Locally this is called *patkar shadi*. In patkar shadi, the wives may be two sisters or may be two different ladies having no relation between them. Two cases have been observed during our survey. In Badkachar village, a man married two sisters and they stay together in the same house. While in Binora village, a man has married twice in two different villages. His wives stay separately in two different huts in the same courtyard.

Alike the Gonds, the Mawasi also practice cross cousin marriage (MBD). The same principles of hypergamy and hypogamy are also found among the Mawasi. Though the Mawasi claim themselves as having equal social status with the Gonds but in practice we found the Gonds are bit superior to the Mawasi. The Gonds never allow anybody to have a Mawasi girl as wife. In that case, the person gets himself ex-communicated. But in Kajri village, one case has been found where a Mawasi man married a Gond girl. The girl has no connection with her parental family due to ex-communication. But the family is living at Kajri village like other Mawasi family happily without any kind of discrimination.

In Mawasi dialect, death is termed as Foth. Body of the deceased is carried to the graveyard (morghat) by the relatives and the villagers on a bamboo stretcher (thatai). Dead body is given a ceremonial bath at burial ground and then covered by a shroud (kappan). In case of male, colour of the shroud is white, while in the case of female it is red. Then the chief mourner (usually son of the deceased) puts kutki rice, ghee, turmeric, betel leaf and water in the mouth of the deceased. While these are being done, others dig a grave in north-south direction. The shrouded corpse is then put inside the grave keeping its head in the southern direction. Everyone offers a fistful of earth into the grave. The grave is then filled up with soil and the loose soil is made compact with the help of spade and stone blocks. They also cover the grave with boulders so that no wild animal can dig into the grave. Everybody then takes bath in the hill stream and returns back to the decease house. Family members of the deceased person offer them cup full of country liquor.

On the third day after *foth*, all the family members again go to morghat along with a few village elders. They carry kutki rice, ghee etc., (as mentioned above) and place those edibles on the grave along with some country liquor. This is known as *teeja*. Final post funeral ritual takes place on the eleventh day, which is locally known as *egarami* or *gatha*. Four days before the ritual, the sons of the deceased person go to every house in their village and put a few grains of rice along with a pinch of turmeric powder on the threshold. They then invite the household members to join in the ritual. On the fixed day, everybody comes to the dead person's house. They take seven small pieces of bamboo, which symbolizes the bones of the deceased. They are washed thoroughly, dried by a piece of cloth and placed on a new piece of cloth and kept over a mud platform in the courtyard. It is known as *jang*. They sacrifice a hen or a goat and its blood is sprinkled over the *jang*. They then put some rice over the blood stained *jang* and sprinkle country liquor. The sacrificed animal is then sent for cooking. The sons of the deceased make a bundle of jang by tying the ends of the cloth and carry that ceremonially to a hill stream or river. Adult male members of the village also follow them in a procession. Then the bundle is thrown into water. It is believed that from there, soul of the dead person becomes free from worldly ties with this ritual and starts its journey to the other world. On their return after the immersion of jang, the communal feast begins. Country liquor is served to all. Then food is served on leaf plates, which contains boiled rice, sag (fried green leafy vegetables), puri (small round shaped wheat bread deep fried in oil), papad (spicy lentil wafers), ghee (clarified butter), meat curry and laddu (sweet meat). They procure a teak plank measuring about 30" in length, 10" in width and 2" in thickness. They scrape its surface with a sharp iron instrument to make wooden plank smooth and even. They then shape it like a flat cricket bat with a very short handle. This wooden plank is called *patia* or *gatha*. Then on one side they start wood carving. They carve out many designs, such as the sun, the moon, horse, human figure (single or multiple) in various poses, name of the deceased persona, and so on. Making of patia takes a long time and whole night they remain awake.

Next morning they rub turmeric paste all over the patia. In a basket they carry patia, gur (molasses), a bottle of country liquor and a coconut. They go to the forest and place the patia ceremoniously under a special tree and worship it with the things mentioned above. Every function is done under the supervision of parihar (priest) and he recites incantations on this occasion. Each got ra (clan) of the village has a separate tree to keep their memorial wooden plank or patia, which may vary according to the convenience of villagers. It may be a banyan or mango or neem, and so on. As per their belief, the deceased soul becomes happy with the offering of memorial plank and even from the other world protects them from any danger or difficulty as an ancestral spirit.

Alike the Gonds, the Mawasis of the PBR core area refer to themselves as Mawasi Thakur. But the neighbouring communities refer them as Mawasi only. Their villages are small and uniethnic. They are located at a lower level between plateaus where several hill streams may be observed. These streams are mostly perennial in nature and act as tributaries for the Denwa and Tawa rivers, which are found inside the PBR core area. Finally they drain out into the Narmada River. Each village has unmetalled and gravely main street, occasionally branched off in different directions. Mawasi houses are situated on either side of the village street either lineally or in clustered fashion, facing the streets. Houses are mostly mud built, thatched with straw, sometimes with polythene sheet and dried grass and in some cases with earthen tiles.

Like the Gonds, there is one multipurpose living room and a patio. Crowding status of the living room is high. If crowding status of a bedroom is above two persons, it is assumed that the privacy of the reproductively active couple gets disturbed. It has been observed that most of the Mawasi families have multiple children of various ages and sexes. The same is true for the Gond families of the core area villages. Doors of the living room are often beautifully carved with different types of designs and coloured with chemical paints available in the markets of Pachmarhi town or Matkuli or elsewhere. During social ceremonies like marriage etc., they decorate the sidewalls of the patio and living room in some cases; they first apply a thick coat of mud with cow dung paste on the wall. While the plaster is wet, they scoop out mud from some selected portions by hand or by a stick or by a sharp iron point. Thus, they make various geometric as well as floral designs on the wall in high relief. Sometimes, reliefs of horse rider, sun, moon, creeper, leaves etc., are also found. They usually let these relief figures to dry up and keep them as they are. In some cases, they colour them with chemical paints. Sometimes, they decorate the sidewalls with different colours of mud, which are locally available. Alike the Gonds, they too have a cattle shed attached to their house. In their homestead, they have kitchen garden and their agricultural fields are also located very near to their house. They use cow dung in the agricultural field as manure. Soil type is sandy and has low water retention capacity. Fields are generally rain fed. Water from hill streams is also used for irrigation where the fields are located beside any stream. All the agricultural fields have fencing around them as protective device from the wild as well as domestic herbivores. Most of the fields also contain watchtower where owner of the field usually stays at night sometimes with his family members.

Mawasi families are mostly nuclear in type. Husband acts as the head of the family. In the absence of husband, the wife takes over the charge till her sons become adult. They do not discriminate between son and daughter, but there exists gender bias among Mawasi parents because only the sons can inherit father's property. In the absence of sons, according to their custom, brother's son takes away property after death. Married girls have no right, except in case of adopted sons-in-law.

Both husband and wife work outside their family and contribute to the family income. In cases of family expenditure, they take joint decision. However, the husband represents his family in the village council meeting. Females are also not allowed to become a mukaddam (village headman) or a parihar (priest).

Mawasi children are trained to accept different responsibilities from very early age, e.g. keeping the homestead clean, grazing domestic cattle in the forest, taking care of the younger siblings, and so on. Girls get gender specific training from their mothers and other elder female relatives of the village. Similarly the boys are trained by their fathers and other elderly males. Mostly at the age of nine or ten years, they accomp any their parents in the jungle and learn the art of foraging.

Similar to, the forest dwelling Gonds, the Mawasis too worship many deities as mentioned below:

1. Kherapati Bai or Kherapati Mata - Village deity, protects from disease and looks after all round welfare of the community

2. Hardul Dada or Hardul Baba	-	Village deity, protects their crops
3. Siddha Baba or Sidhmaraj	-	Village deity, protects water sources in and around the village
4. Kurupan Bai or Kurupan Devi	-	Village deity, looks after the welfare of the village cattle
5. Dongar Deo	-	Hill god, protects them in the high hills during honey collection and other related purposes
6. Bajrang Dada	-	Monkey god, protects them from all hazards
7. Doval Deo	-	Village deity, looks after family welfare
8. Balkhan Deo	-	Forest deity, protects people in the forest
9. Balkunwar Deo	-	Forest deity, protects people in the forest
10. Badin Deo	-	Forest deity, protects people in the forest
11. Keminte Devi	-	Forest deity, protects people in the forest
12. Mailee Devi	-	Forest deity, protects from epidemic

Besides, they also participate in many Hindu festivals of this region, such as, Dussera, Diwali, Maha Shivratri and Nagpanchami.

Present Condition of the Core Area Villages

In the PBR core area, there are two types of villages, viz. (i) the revenue villages and (ii) the forest villages. The revenue villages happen to be the old forest settlements. When the British declared these areas as 'reserved forest', they imposed many restrictions on the age-old foraging system of these villages and at the same time made a thorough survey of these settlements and imposed tax on them to collect revenue. Since then, these settlements have been designated as the revenue villages. No non-tribal can purchase land in these villages as they are situated inside the reserved forest area. They have well demarcated boundaries and also are permanent villages. On the other hand, the forest villages are temporary villages with no fixed boundary. They were created by the colonial forest administration to get continuous labour supply in the extraction of timber from the forests. They mostly employed local tribes for this purpose. During colonial administration, forests of these areas were divided into many blocks and systematic extraction of timber was done from these blocks. Forest villages were constructed near the forest blocks from where they wanted to extract timber on some vacant space. Along with the logging, the labours were allowed to grow crops in the vacant forest areas nearer to their huts and raise cattle. But they had no ownership rights (patta) on these lands. When logging activities finished in one block, the village had to shift to a new area in another block. There was some kind of transience, which has been stopped abruptly after independence. This caused a very bad impact on the socio-economic life of the forest villagers. As a natural outcome, it affected the prevailing bio-diversity of their surrounding areas adversely and goes against the principles of a so-called 'biosphere reserve'.

Similar to other tribal areas of India in this area too, apart from homestead lands and some permanent cultivation fields, all the natural resources belong to the villages. They have community property rights (CPR) over those natural resources. Their utilization is regulated by the village council, e.g. selection and distribution of shifting cultivation plots among the families, burial ground, grazing ground, shrines of the village deities, and so on. Since 1977, due to various governmental regulations and acts, there has been significant erosion in their traditional communal property rights.

The core area of PBR contains nine human settlements. These are, (i) Bariaam, (ii) Jambudweep, (iii) Pagara Dhana, (iv) Ghoranar, (v) Badkachar, (vi) Rorighat, (vii) Kajri, (viii) Binora, and (ix) Pattan (Chittor Pattan). Among them, Ghoranar is a forest village and the rest are revenue villages. The following table (Table-I) will show about the loss of communal property rights among the above-mentioned 'core area' villages.

Total population in the 'core area' villages' during the present survey was 1451 (Male-745, Female-706), with sex ratio being 948 females per thousand males. Village wise distribution of population along with their community, age and sex has been given in the following table (Table-1.2).

Table 1.1: Loss of community property right in the 'core area' villages in PBR

Sl. No.	*Name of the Villages*	*Total Area (in Htrs.) before 1977*	*Present area of the village (Htrs.)*	*% of loss in CPRs*
1.	Bariaam	4977.420	110.698	97.78
2.	Jambudweep	180.901	1.986	98.90
3.	Pagara Dhana	2161.295	29.547	98.63
4.	Ghoranar	-	-	-
5.	Badkachar	1854.420	41.005	97.79
6.	Rorighat	829.609	214.891	74.09
7.	Kajri	1005.754	14.629	98.54
8.	Binora	764.202	61.697	91.92
9.	Pattan	-	-	-

Source: Collected from Satpura Tiger Reserve Project Office, Pachmarhi vide Forest Department Notification, Government of Madhya Pradesh, No. F-15/22/76-10(8); dated 01-06-1977 (C.P.R= Community Property Right)

Table 1.2: Age, sex and community-wise distribution of population in the 'core area' villages of PBR

Village	*Community*	*0-5*		*6-10*		*11-15*		*16-20*		*21-25*		*26-30*		*31-35*	
		M	*F*	*M*	*F*	*M*	*F*	*M*	*F*	*M*	*F*	*M*	*F*	*M*	*F*
Bariaam	Gond	-	2	3	3	6	6	6	5	1	2	-	3	1	4
- do -	Mawasi	-	-	-	-	2	-	-	-	-	-	-	1	1	-
- do -	Rajput	2	-	-	-	-	-	1	-	-	1	1	-	-	1
- do -	Yadav	1	-	-	1	1	-	2	1	-	-	-	1	1	1
- do -	Muslim	-	-	-	-	-	-	-	-	-	-	-	1	-	-
Jambudweep	Yadav	3	2	3	4	9	6	9	6	1	3	1	6	3	6
		-	-	-	-	-	-	1	1	1	-	1	-	-	-
Pagara Dhana	Mawasi	12	14	8	14	7	7	8	1	2	7	4	4	5	5
- do -	Gond	1	1	-	2	-	2	-	1	1	-	-	1	1	-
Ghoranar	Mawasi	13	15	8	16	7	9	8	2	3	7	4	5	6	5
		8	14	14	8	10	6	5	6	10	6	2	2	2	1
Badkachar	Mawasi	17	16	18	17	12	10	8	12	9	13	9	4	3	6
Rorighat	Mawasi	25	21	18	25	14	14	14	13	13	11	11	10	10	7
Kajri	Mawasi	24	19	16	20	13	7	8	7	11	10	6	5	2	4
- do -	Gond	3	-	-	2	-	1	1	-	-	-	-	-	1	1
Binora	Mawasi	27	19	16	22	13	8	9	7	11	10	6	5	3	5
		26	20	23	21	13	22	13	17	9	8	6	4	7	10
Pattan	Gond	24	16	17	10	15	7	5	10	17	21	15	10	11	5
TOTAL		143	123	117	123	93	82	72	74	74	79	55	46	45	45

(Contd.)

(Contd. Table : 1.2)

Village	Community	36-40		41-45		46-50		51-55		56-60		61 +		Total		
		M	F	M	F	M	F	M	F	M	F	M	F	M	F	T
Bariaam	Gond	2	5	4	1	2	1	-	-	1	1	-	1	26	34	60
- do -	Mawasi	-	-	-	-	-	-	1	-	-	-	-	-	4	1	5
- do -	Rajput	-	-	1	-	-	-	-	-	-	-	-	-	5	2	7
- do -	Yadav	-	-	1	-	-	-	-	-	-	-	-	1	6	5	11
- do -	Muslim	1	-	-	-	-	-	-	-	-	-	-	1	1	2	3
Jambudweep	Yadav	3	5	6	1	2	1	1	-	1	1	-	3	42	44	86
		-	2	1	-	1	-	-	-	-	-	-	-	5	3	8
Pagara Dhana	Mawasi	4	4	2	1	2	1	-	-	-	2	2	1	56	61	117
- do -	Gond	-	-	-	-	-	-	-	-	-	-	-	-	3	7	10
Ghoranar	Mawasi	4	4	2	1	2	1	-	-	-	2	2	1	59	68	127
		1	5	4	2	4	1	1	1	1	-	-	-	62	52	114
Badkachar	Mawasi	4	5	6	3	2	4	4	3	1	3	4	4	97	100	197
Rorighat	Mawasi	7	8	4	6	8	6	2	1	4	3	-	1	130	126	256
Kajri	Mawasi	7	4	4	4	1	-	1	1	1	1	-	2	94	84	178
- do -	Gond	-	-	-	-	-	-	-	-	-	-	-	-	5	4	9
Binora	Mawasi	7	4	4	4	1	-	1	1	1	1	-	2	99	88	187
		8	5	6	5	3	2	2	1	1	2	2	3	119	120	239
Pattan	Gond	5	5	5	5	6	4	3	3	2	2	7	7	132	105	237
TOTAL		39	43	38	27	29	19	14	10	11	14	15	21	745	706	1451

Bariaam Village:

Bariaam is situated about 9 km away from Pachmarhi town in the northeastern direction. It is a multiethnic revenue village; having 21 families (Gond-15, Mawasi-2, Rajput-2, Yadav-1 and Muslim-1); inhabited by 86 individuals. Average family size of the village is 4.1, though different communities show different averages (e.g. Gond-4.0, Mawasi-2.5, Rajput-3.5, Muslim-3.0 and Yadav-11.0).

From higher plane, this village looks like an upturned bowl, surrounded by Bariaam Lake, Pachmarhi-Pipariya road and forests of Pachmarhi Sanctuary. Houses are constructed on both sides of the village street in a linear fashion with well-demarcated boundaries (vide Map 1).

It is reported that the Gonds were the original settlers; the other communities came to settle there at a later time. This village has three uninhabited brick built buildings, with modern electrical appliances such as geyser, ceiling, and so on. It was reported that famous author-cum-social activist, Ms. Arundhati Ray of *'God of Small Things*' fame happens to be one of the owners. There is a restriction that no non-tribal can buy land in the tribal villages located inside the reserved forest. Even though, those were built, taking sufficient time for being constructed. It is surprising that the state forest authority kept their eyes shut till those were completed. However, the local forest officials did not answer to our queries that this matter was sub judice. A similar thing was found at Madai on the way to Pattan village, which will be discussed later on.

Village huts are mostly mud built, thatched with earthen tiles (khapra). There are a few brick built houses, which were constructed with state governments financial assistance before this village was declared as a part of protected area. Each house opens to the village street and has a sacred basil (tulsi) plant in their courtyard (angan). In some huts lower part of the sidewalls are painted with chemical paints. Most of the huts have cowshed (sar) attached with them. These have wooden or bamboo sidewalls thatched with straw or dried grass. A few houses have covered latrines and others go to forest for answering nature's call. They collect drinking water from the village well. A tube well is also there. It is out of order for the last couple of years. For bathing, washing

Map-1

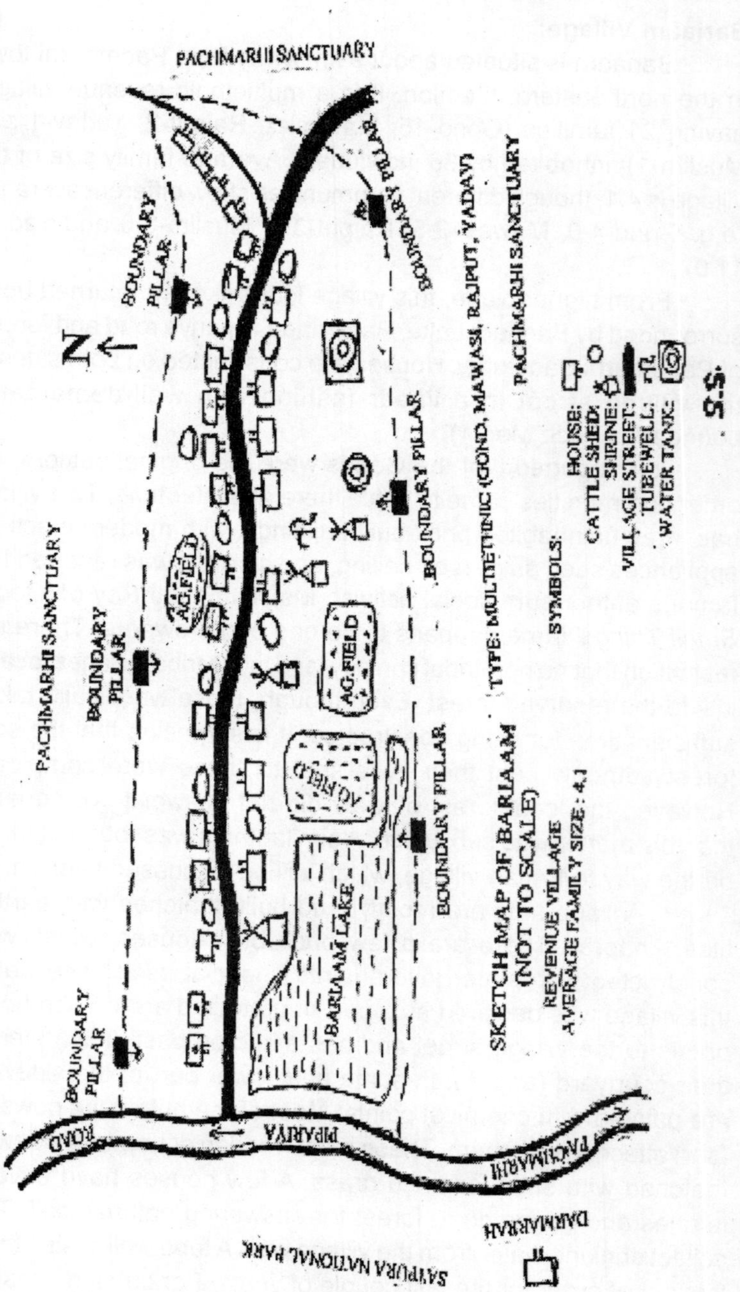
PACHMARHI SANCTUARY
PACHMARHI SANCTUARY
PACHMARHI SANCTUARY
BOUNDARY PILLAR
BOUNDARY PILLAR
BOUNDARY PILLAR
BOUNDARY PILLAR
BOUNDARY PILLAR
BOUNDARY PILLAR
BOUNDARY PILLAR
N
AG. FIELD
AG. FIELD
AG. FIELD
BARIAAM LAKE
SKETCH MAP OF BARIAAM
(NOT TO SCALE)
REVENUE VILLAGE
AVERAGE FAMILY SIZE : 4.1
TYPE: MULTIETHNIC (GOND, MAWASI, RAJPUT, YADAV)
SYMBOLS:
HOUSE:
CATTLE SHED:
SHRINE:
VILLAGE STREET:
TUBEWELL:
WATER TANK:
S.S.
PIPARIYA
ROAD
PACHMARHI
DARMARRAH
SATPURA NATIONAL PARK

and other household purposes they fetch water from the Bariaam Lake. This village has electric connection. Five families have television setts (out of them one is coloured television) and one family has Direct to Home (DTH) antenna installed over its roof.

Before 1977, Bariaam had 4977.420 hectares of land. It included habitation, cultivation and forestlands along with lake. At present this village possesses 110.698 hectares of land. It means that the state forest department has taken away about 97.78% of total village land in 1977 without giving any sort of compensation. Due to lack of sufficient cultivable land, agriculture is not their primary occupation. They depend mainly on forest for collection of firewood and some other non-timber forest produces (NTFP). Villagers reported that they usually collect mango (mangifera indica) and achar or chironji (buchnania latifolia) from forest and sell them to the agents at Pachmarhi town. They declined to report about the other produces, as there is forest department's restriction on their collection. Even though their poverty compels them to collect those in small quantities and sell them to the agents at Pachmarhi or Matkuli. They fetch good price. There are herbal medicine shops at Pachmarhi and other neighbouring urban centres, which depend on the tribals for the supply of medicinal herbs and plants. Though collected in small quantities but ever increasing number of collectors may be exerting great pressure on the indigenous medicinal herbs and plants. In Bariaam, 58 individuals (18-60 years) of both sexes are involved in collection of firewood and NTFPs. Each person brings a head load of firewood weighing about 20 to 25 kilograms on every alternate day. Selling price of each head load varies from Rs. 20/- to Rs. 25/- in Pachmarhi market. Collection of firewood is done in a team. Either husband and wife, or village females or village male folk make teams for foraging. A team may have 2 to 6 persons.

Besides foraging, they also work as day labourers as and when they get opportunity for that. Daily wage is negotiable and varies from Rs. 30/- to Rs. 50/- per day per head depending upon the nature of job. On an average they get less than seven days employment as daily wage labourer. They occasionally work for State Public Works Department during road repairing and for State Forest Department for the clearance of 'lantana bush' inside the

Satpura National Park and Pachmarhi Sanctuary. During festivals, such as Maha Shivratri and Nagpanchami, when thousands of devotees come on pilgrimage, they go to work in the hotels and shops at Pachmarhi town as daily wage labourer. Every year in May-June, they go to the plain villages around Pipariya to work as agricultural labour during the harvesting of wheat, soybean, and so on. Soil in Bariaam is sandy (retily). They use it as kitchen garden and grow tomato, chilli, egg plant, flat beans, radish, etc., for their own consumption.

Bariaam has a primary school without any building. Classes are held on the covered patio of a house, where a single teacher coaches students from Class-I to Class-IV. Village boys and girls go to Pachmarhi town for secondary education.

Pagara Dhana Village:

Pagara Dhana is a multiethnic revenue village, situated about 16 Km away from Pachmarhi town in the northern direction. It is situated inside the Satpura National Park (vide Map 2).

For moving inside village one has hill streams at several places. This village shows a scattered settlement pattern. It has 23 families out of which 2 families are Gond and the rest 21 are Mawasi families. Average family size is 5.5. Before 1977, Pagara Dhana possessed 2161.295 hectares land. At present it has 29.547 hectares of land, i.e. about 98.63% of village land has been taken away by the Satpura National Park Authority without paying any compensation to the villagers and has erected pillars (munara) on all sides so that village settlement remains confined inside the marked area only.

Each family lives in mud huts with attached cowshed (sar). Boundary of each house has fencing. Sidewalls of the mud huts are made of split bamboo strips or long dried branches of tree and plastered on both sides with a thick paste of mud and cow dung. Houses have rectangular ground plan, divided into one main living room and a patio (may be opened or closed on two sides). Main room is partially divided for kitchen and storeroom where grains are stored in big bamboo baskets plastered on all sides by a paste of mud and cow dung. Inside the living room they use string cots

Map - 2

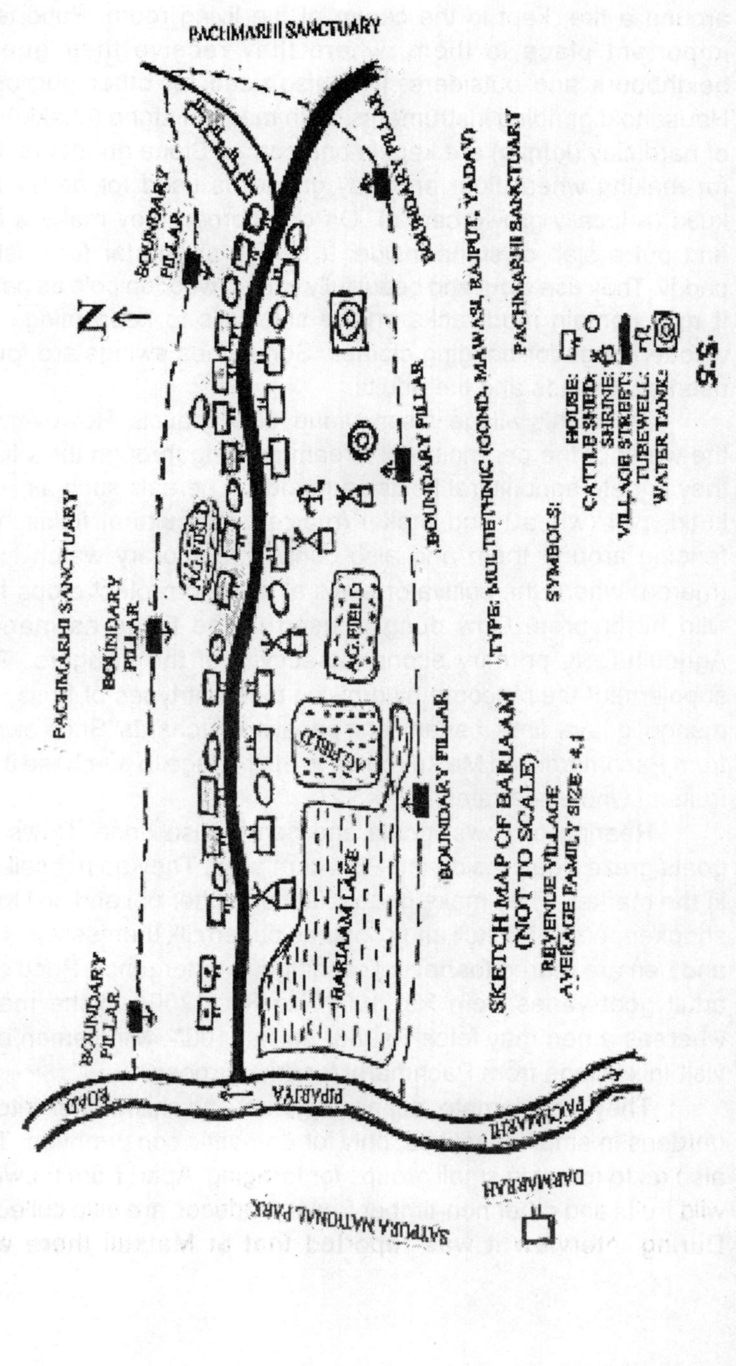
PACHMARHI SANCTUARY
PACHMARHI SANCTUARY
PACHMARHI SANCTUARY
BOUNDARY PILLAR
BOUNDARY PILLAR
BOUNDARY PILLAR
BOUNDARY PILLAR
BOUNDARY PILLAR
BOUNDARY PILLAR
BOUNDARY PILLAR
N
AG. FIELD
AG. FIELD
AG. FIELD
BARIAAM LAKE
SKETCH MAP OF BARIAAM
(NOT TO SCALE)
REVENUE VILLAGE
AVERAGE FAMILY SIZE : 4.1
TYPE: MULTIETHNIC (GOND, MAWASI, RAJPUT, YADAV)
SYMBOLS:
HOUSE:
CATTLE SHED:
SHRINE:
VILLAGE STREET:
TUBEWELL:
WATER TANK:
ROAD
PIPARIYA
PACHMARHI
SATPURA NATIONAL PARK

(charpai) for sleeping but majority of the people sleep on the floor around a fire, kept in the centre of the living room. Patio is an important place to them, where they receive their guests, neighbours and outsiders. It is also used for other purposes. Household grinding instruments, both made of stone (chakki) and of hard clay (kunita) are kept in one corner. Stone grinder is used for making wheat flour and clay grinder is used for de-husking kutki (a locally grown cereal). On one corner, they make a hole and put a slab of stone inside. It serves as mortar for husking paddy. They use stout and beautifully carved wooden pole as pastel. It may contain mud racks on the sidewalls to keep things and wooden pegs for hanging clothes. Sometimes swings are found, used by the kids and the adults.

Soil of this village is sandy and full of gravels. However with the water of the perennial hill stream passing through the village, they irrigate agricultural fields and produce cereals such as kodo, kutki, pisi (wheat) and makai (maize). Agricultural fields have fencing around them and also contain temporary watch tower (maroa) where the cultivator stays at night to protect crops from wild herbivores. Cow dung is used in the fields as manure. Agriculture is primary economic activity of the villagers. They supplement their income by growing different types of fruits, viz., mango, guava, lime, banana and papaya in orchards. Shop owners from Pachmarhi and Matkuli come to their village to purchase these fruits at wholesale rates.

Rearing of cows, goats and hen is also done. Cows and goats graze freely inside the PBR core area. They do not sell milk in the market. They make ghee (clarified butter oil) and sell to the shopkeepers at Matkuli and consume buttermilk themselves. Goat and hen are reared for having cash during emergency. Price of an adult goat varies from Rs. 1000/- to Rs. 1200/- in the market whereas a hen may fetch Rs. 80/- to Rs. 100/-. Middlemen often visit this village from Pachmarhi for this purpose.

They grow tomato, beans, eggplant and chilli in their kitchen gardens in small quantities, only for domestic consumption. They also go to forest in small groups for foraging. Apart from firewood, wild fruits and other non-timber forest produces are also collected. During interview it was reported that at Matkuli there were individuals of Sahoo (businessman) caste. They are the villagers best friend. They purchase their collections. If they require anything they send their messenger to the village. Sometimes they even lend them money as advance. This relationship is very old and continuing for many generations. However, the villagers were reluctant to tell the names NTFPs, except mango and achar.

In reality Pagara Dhana is a dhana (hamlet) of the Pagara village, which is situated on the eastern side of the Pachmarhi-Pipariya road. This village has no electricity. This is a malaria endemic zone. People collect drinking water both from tube well and perennial hill streams. The village shrine, primary school, Anganwadi Centre etc., is located in the main village.

Ghoranar Village:

Ghoranar is an uniethnic forest village, situated on the western direction of Pagara Dhana village, about 22 Km away from Pachmarhi town. It contains 17 Mawasi families. Average family size is about 6.7. According to forest department's information 'forest villages' have no well-defined boundaries like 'revenue villages' and total area belongs to the forest department. During survey from the pillars (munara) erected by the State Forest Department, it appeared that while being a 'forest village', Ghoranar has well defined boundary (vide Map 3).

It shows linear settlement pattern. Houses are constructed on both sides of the village street.

Ghoranar is a flat village and its northern and southern boundaries are rocky. It has two perennial hill streams on its two sides. Due to rocky terrain, water from these streams cannot be diverted towards agricultural fields. All the agricultural fields are rain fed. In comparison to Pagara Dhana village, agricultural as well as kitchen garden yields of Ghoranar are less. Here the villagers depend mostly on foraging, rearing of domestic cattle and poultry birds (as reported earlier in case of Pagara Dhana village). Agricultural fields are protected on all sides by fencing. Fences are made up of thorny branches of trees and bamboo. Local term for the protected cultivation plot is *bari*. At one corner of bari they erect a straight wooden or bamboo pole. On top of it, they tie bunches of maize corns serially in layers, locally called as

Map - 3

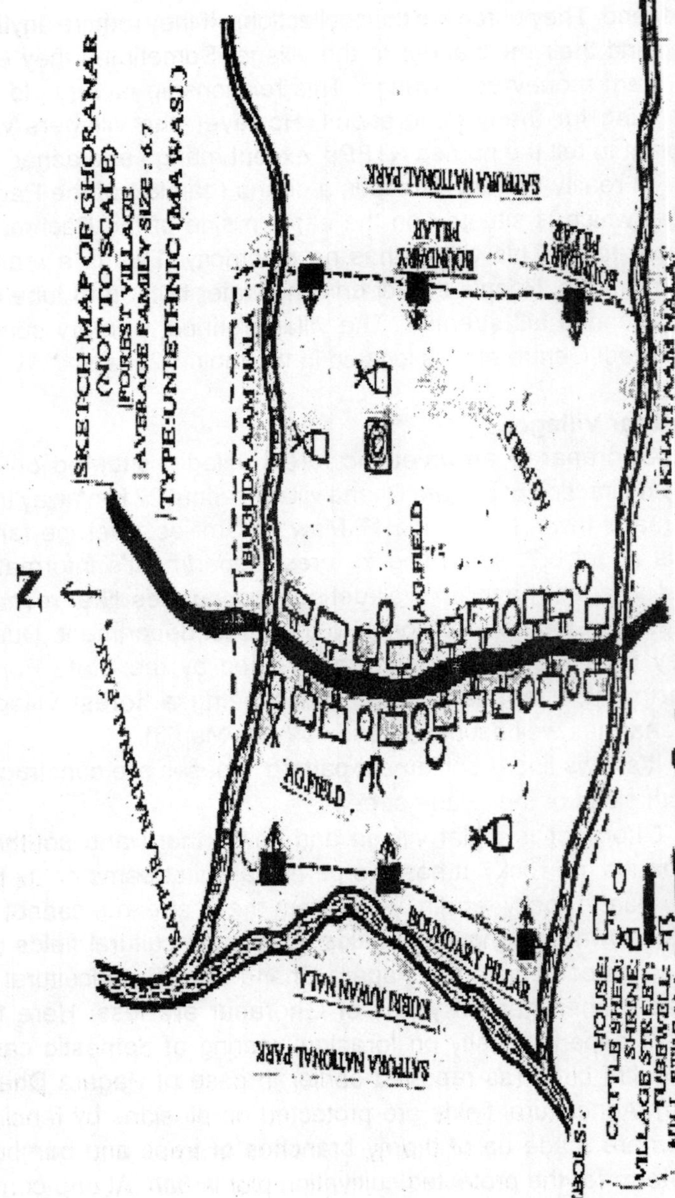
SKETCH MAP OF GHORANAR
(NOT TO SCALE)
FOREST VILLAGE
AVERAGE FAMILY SIZE : 6.7
TYPE:UNIETHNIC (MAWASI)
N
SATPURA NATIONAL PARK
BOUNDARY PILLAR
AG.FIELD
KHATIAAM NALA
SATPURA NATIONAL PARK
S.S.
SYMBOLS:
HOUSE:
CATTLE SHED:
VILLAGE SHRINE:
VILLAGE STREET:
TUBEWELL:
HILL STREAM:
WATER TANK:

jera. It looks like a huge bottle cleaning brush. Top of the jera are covered with thick layers of dried grass or any other waterproof material so that rainwater may not trickle down and damage corns. Inside their houses also they preserve maize by tying bunches of them in a bamboo or wooden pole. However, other cereals like *kutki*, *kodo* etc., are stored inside the living room in mud plastered bamboo baskets.

Village thoroughfare is wide and being used in type of social occasion or ceremonies. It also serves as playground of the village kids. Sidewalls of the village huts are mostly made of split bamboo or by long slender branches of tree, plastered on both sides with thick paste of mud and cow dung. In a few cases, it was observed that stone blocks have been used for the construction of sidewalls where mud was used both as cementing and plastering material. They erect several strong and stout wooden poles on the floor. Over these poles and the sidewalls, they make a frame either by bamboo or by slender and round branches of trees. This frame serves as the infrastructure of roof and is thatched either by earthen tiles or by dry long grasses collected from jungle or by straw. As there is restriction on the collection of long grasses from the jungle, they also use long thin plastic sheets along with grass. As stated earlier, each house has a multipurpose patio on its front side. Sidewalls of these patios in some cases are decorated with carved geometric designs. Stylized carved figures of sunflower, lotus, creepers, horse rider etc., are also found there. Apart from agriculture, making of ghee (clarified butter/butter oil), adult individuals of both sexes also go out to work as daily wage labourer (as reported in the case of Bariaam village). They leave village very early in the morning. Eldest child looks after his/her younger siblings throughout the day. Sometimes very old ladies look after the village children.

Like Bariaam village, Ghoranar too has a primary school, without any school building where a single teacher coaches the students (Class I to V). Most of the village boys and girls attend school regularly as it provides 'midday meals'. 'Midday meal' consist of two thin handmade chapattis (thin flat wheat bread) and a little vegetable.

Due to economic hardships and other objective conditions, the Mawasi children of Ghoranar village get practical training in mutual help and co-operation at a very early age and it helps to develop a permanent emotional bond of fellow feeling among them.

Badkachar Village:

Badkachar is a uniethnic village, inhabited by the Mawasi people, situated about 16 Km away from Pachmarhi town in the northern direction. However, the villagers' do not use this route. They usually walk down through a very narrow shorter forest route over a small hill to reach Pachmarhi. Village border is well demarcated by pillars (munara) and it is surrounded by the Satpura National Park on all sides (vide Map 3).

It has 35 families. Average family size is 5.6. It is almost a plain area encircled by hills and has a beautiful landscape.

At the entrance of the village stands a huge banyan tree under which the shrines of Bajrangdada or Hanumandada and Sidh Maharaj are situated. All the families live in mud built huts, which face the uneven, gravelly unmetalled and serpentine village street. Arrangement of huts exhibit scattered clustered settlement pattern. This village had 1854.420 hectares of land before 1977, i.e., before the declaration of Pachmarhi Sanctuary. At present, the village has 41.005 hectares of land in its possession. It means the Government of Madhya Pradesh has taken away 97.79% of their community property without any compensation. Though they try hard to remain within the new village boundary even then some of their agricultural fields have been found beyond the boundary pillars. A hill stream flows along the entire southern boundary of the village then bends towards east and covering the eastern side flows towards Pagara Dhana village. Agriculture is primary occupation of the villagers. They use stream water for irrigation. There are two wells also in the village. Adjoining fields of these wells water for the said purpose. Previously villagers used to take out water with the help of pulley and bucket. Now some of them use diesel pump set for the same. The other fields located in the central and northern part of the villages are rain fed. They grow wheat, *kodo*, *kutki*, gram and potato. They also grow eggplant, flat

Map - 4

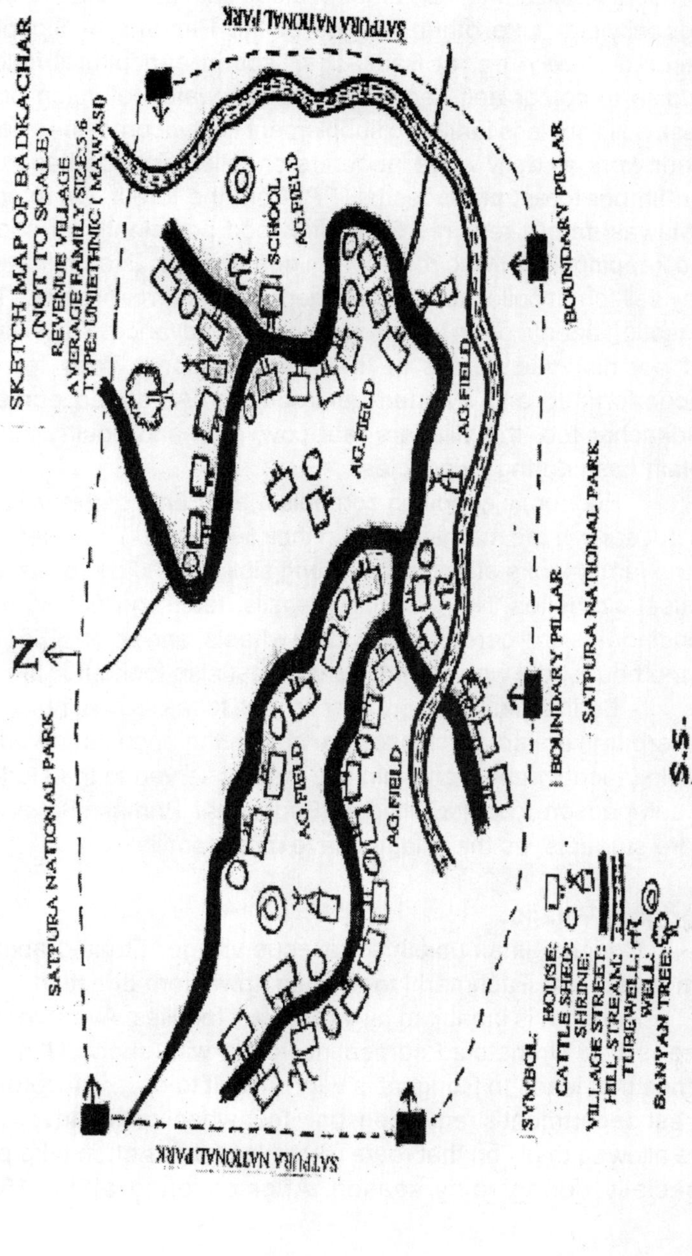
SKETCH MAP OF BADKACHAR
(NOT TO SCALE)
REVENUE VILLAGE
AVERAGE FAMILY SIZE:5.6
TYPE: UNIETHNIC (MAWASI)
N
SATPURA NATIONAL PARK
SATPURA NATIONAL PARK
SATPURA NATIONAL PARK
SATPURA NATIONAL PARK
BOUNDARY PILLAR
BOUNDARY PILLAR
SCHOOL
AG.FIELD
AG.FIELD
AG.FIELD
AG.FIELD
AG.FIELD
SYMBOL:
HOUSE:
CATTLE SHED:
SHRINE:
VILLAGE STREET:
HILL STREAM:
TUBEWELL:
WELL:
BANYAN TREE:
S.S.

beans, chilly and a few leafy vegetables in their kitchen garden. Cow dung is extensively used as manure. A few of them also use chemical fertilizer in small amount along with pesticides to obtain a good yield. Like other villages of the Pachmarhi Biosphere Reserve Core Area, at Badkachar, soil in agricultural fields is reddish in colour and sandy. Due to low water retention power, these soils are less fertile. To supplement their income, the villagers either work as daily wage labourers or collect firewood and other non-timber forest produces (NTFP) from the forest. On average, a Mawasi family requires 50 kg firewood per month for cooking and keeping their living room warm during night. Except firewood, they sell forest collections to the agents at Pachmarhi town. These collection agents often lend them money in advance. The villagers did not disclose details of these transactions. They felt very uncomfortable and avoided our requests. As stated earlier, at Badkachar too, the villagers rear cow, goat and poultry birds to obtain cash during exigencies.

Economic condition compels the parents of this village to go out early in the morning to earn their livelihood. The eldest child in the family looks after his/her young siblings and performs some household duties, like cleaning utensils, fetching drinking water, grinding/husking cereals by grinding wheels, and so on. Old ladies of the household who cannot go to forest also look after the kids.

Badkachar has a primary school. It has its own hut. As the classrooms are small, classes are held in the open courtyard. It is single teacher school and mid day meal is served to the students. In comparison to other villages, Badkachar Primary School has more students, as the villages have more families.

Rorighat Village:

Rorighat is an uniethnic revenue village, situated about 18 Km away from Pachmarhi town in southwestern direction.

Rorighat is inhabited by 44 Mawasi families. Average family size is 5.8. Unmetaled serpentine roads with sharp 'U' shaped bends that leads to Rorighat is very difficult to cover. According to forest department's regulation, only four wheel petrol driven jeeps are allowed to ply on that route where landslides often take place, specially during rainy season. After covering about 15 km,

Map - 5

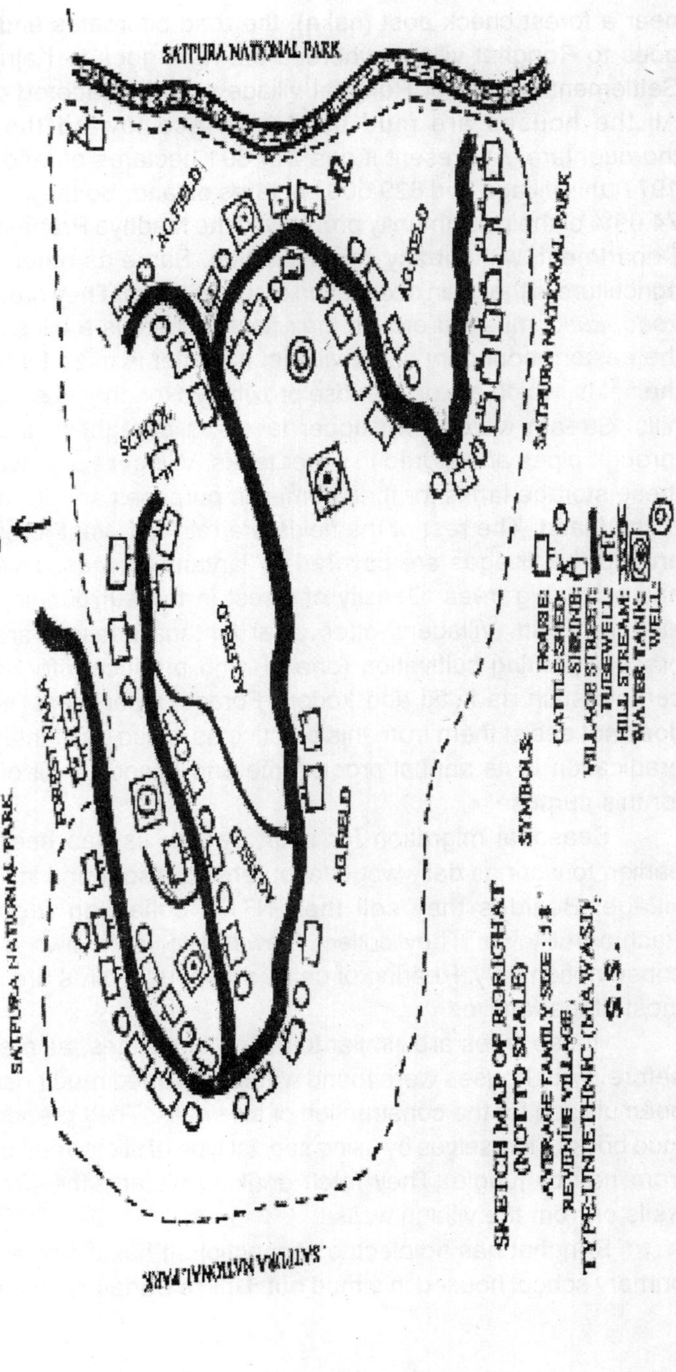
SATPURA NATIONAL PARK
SATPURA NATIONAL PARK
SATPURA NATIONAL PARK
SATPURA NATIONAL PARK
N
SCHOOL
FOREST NAKA
AG. FIELD
AG. FIELD
AG. FIELD
AG. FIELD
AG. FIELD
AG. FIELD
SYMBOLS
HOUSE:
CATTLE SHED:
SHRINE:
VILLAGE STREETS:
TUBEWELL:
HILL STREAM:
WATER TANK:
WELL:
SKETCH MAP OF RORIGHAT
(NOT TO SCALE)
AVERAGE FAMILY SIZE : 3.8
REVENUE VILLAGE
TYPE-UNIETHNIC (MAWASI)
S.S.

near a forest check post (naka), the road bifurcates and the left goes to Rorighat village whereas the right goes to Kajri village. Settlement pattern of Rorighat village exhibit scattered clusters. All the houses are mud build and face toward the village thoroughfare. At present it has 214.891 hectares of land. Before 1977, this village had 829.609 hectares of land. So they lost about 74.09% of their community property to the Madhya Pradesh Forest Department, without any compensation. Same as other villages, agriculture is the main occupation of the villagers. They grow wheat, *kodo, kutki*, mustard etc., in their fields. There is a hill stream on the eastern boundary of the village. Its water is used for irrigating the fields, which are on its close proximity. Rorighat is encircled by hills. Stream water from upper levels is brought to the village through pipes and stored in water tanks. Villagers use water from these storage tanks for their domestic purposes and for irrigation of their land. The rest of the fields are rain fed. Vast forest areas around the villages are covered by lantana bushes. This village has a few big trees. Density of forest in the surrounding hills is also very thin. Villagers' often clear lantana infested areas and practice shifting cultivation (dhaiya) and produce inferior quality cereals such as kutki and kodon. Forest Department, however, does not desist them from this practice as it also has lantana bush eradication in its annual programme and spends a lot of money for this purpose.

Seasonal migration for short period, as has been stated earlier, to work as daily wage labourerer is also found in Rorighat village. Besides they sell their NTFP collection regularly in Pachmarhi town. They collect firewood for their own domestic consumption only. Rearing of cattle and poultry birds are done by most of the families.

House types are similar to the other villages, as mentioned before. Four houses were found where sun dried mud bricks have been utilized for the construction of sidewalls. They prepare these mud bricks themselves by using special type of sticky mud collected from nearby jungle. They fetch drinking water either from tube wells or from the village wells.

Rorighat has no electric connection. It has a single teacher primary school housed in a mud hut. Unlike Bariaam village, every house in this village has grinding stones, mortar and pastels. It signifies that they manually process their grains for consumption. During survey, it was observed that the villagers use bullocks for thrashing their harvest.

Kajri Village:

Kajri is a multiethnic revenue village, situated about 25 km away from Pachmarhi town in the western direction. Road leading to Kajri is very stiff and difficult, full of loose gravels and boulders of different sizes. It is a bowl shaped valley inside the PBR core area, surrounded by hills on all sides such as the Nishangarh in the east, the Brijlaldev Hills in the west, the Dhotigarh in south and the famous Nagdwari Hills in the north. It has one Gond and 28 Mawasi families. Average family size is 6.4. Settlement pattern is scattered cluster.

Though it is a very difficult and remote village from the point of view of communication facilities, yet Kajri is a very important pilgrimage centre. Every year during Nagpanchami festival (July-August), about 200,000 to 250,000 pilgrims pass through this village. All the village houses are mud built, thatched with either earthen tiles or straw.

Each house has covered patio (osari) in front. Size of the patios usually varies from 90 sq. ft. to 350 sq. ft. They maintain these patios carefully, sometimes paint them with chemical hues and put a quartz wall clock on the sidewall. Moreover, there may be farmed coloured photographs of the gurus of different ashrams, whose disciples' visit there. Nagpanchami festival lasts for two weeks only. But preliminary agreements start well before that. Various reputed ashrams of Maharashtra take part in this festival. Authorities of these ashrams come well in advance to erect their camps in the village. They inform the villagers about their arrival date at Pachmarhi town and the number of porters they require because they bring all food materials, oil, etc., not only for themselves but also for free distribution of cooked food among the devotees. Youths of Kajri village bring those materials as head load on foot from Pachmarhi to Kajri @ Rs. 100/- per head per day. The festival is observed during rainy season. Dharamsalas (free guest houses for pilgrims) do not have sufficient

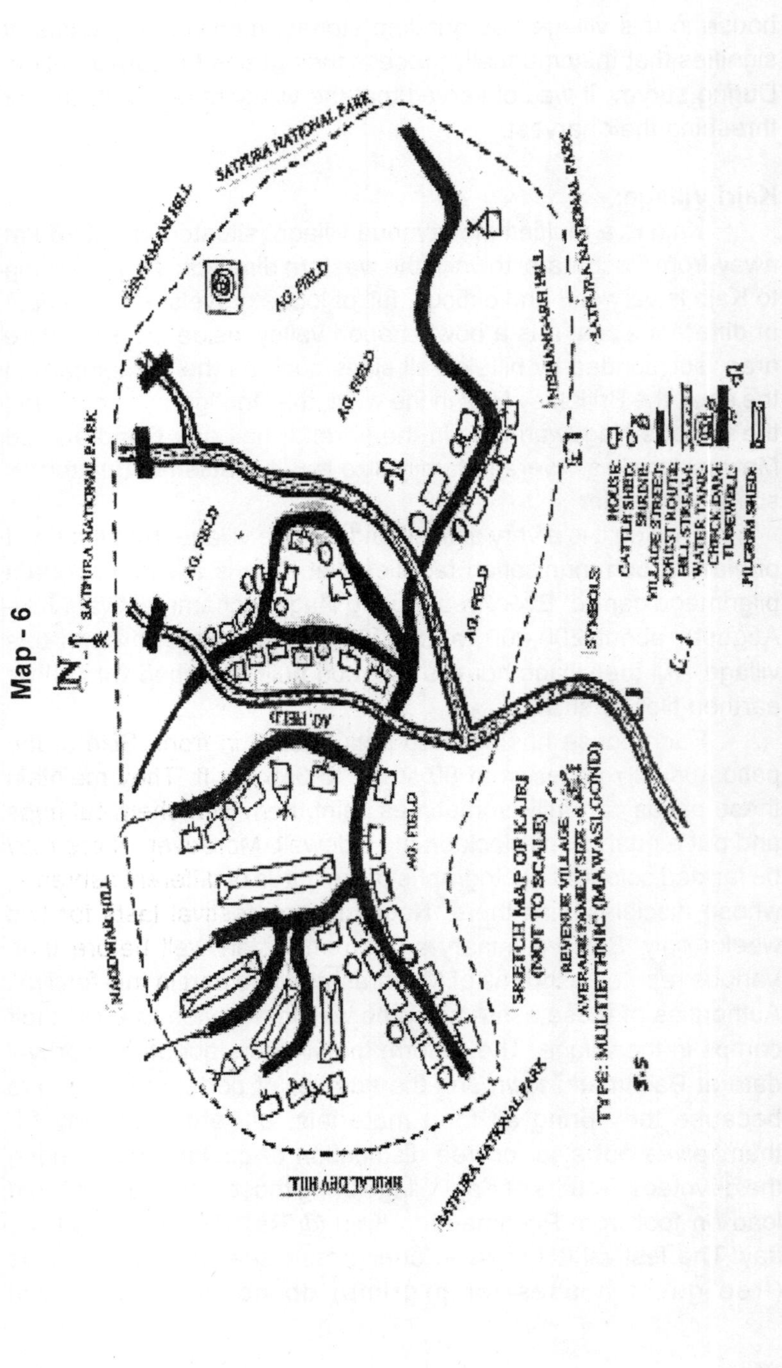
Map - 6
N
SATPURA NATIONAL PARK
CHINTAMANI HILL
NAGDWAR HILL
BHILAL DEV HILL
NISHANGARH HILL
AG. FIELD
SKETCH MAP OF KAJRI
(NOT TO SCALE)
REVENUE VILLAGE
TYPE: MULTIETHNIC (MAWASI,GOND)
SYMBOLS
HOUSE
CATTLE SHED
SHRINE
VILLAGE STREET
FOREST ROUTE
HILL STREAM
WATER TANK
CHECK DAM
TUBEWELL
PILGRIM SHED

accommodation. So the pilgrims stay on the covered patios in the village. Rent of the sleeping space varies from Rs. 15/- to Rs. 30/- per person per night. The following Figure 1 is a general ground plan of a typical Mawasi hut in Kajri.

Not only *osari* and *pichhore,* during the peak period the villagers even rent out part of their living room also to the pilgrims. On an average depending on the size of their patios, earnings of each family vary form Rs. 300/- to Rs. 500/- per day. They also sell firewood to the pilgrims (@ Rs. 10/- per bundle of 5 kg) who want to prepare their food themselves. The house owner also brings drinking water, etc., for the pilgrims. He gets a portion of food from the pilgrims in return as tips. So for the entire period of Nagpanchami festival, the villagers get free food either from the pilgrims or from the ashrams for which they work on daily wage basis. While leaving village, the pilgrims also give uncooked rice, wheat flour, cooking oil, etc., to the house owner, which remained unconsumed. These items, thus collected, provide food for each family for about 3 to 4 months. Similarly the ashram authorities also distribute their unspent supplies among the villagers to maintain good public relations. During festival time the villagers also set up various temporary shops for the pilgrims. Apart from honey, tea, mohua liquor, betel leaf, cigarettes, chewing tobacco packets, etc., they also sell candles, vermilion powder and sugar wafers to the pilgrims. They charge Rs. 100/- for a litre of honey-collected from the forest, Rs. 2/- for a cup of tea, Rs. 5/- for a glass (about 100 ml) mohua liquor or Rs. 40/- for a bottle of one litre and sell the other items with a little premium than the printed prices on their packs. There are two distinct routes for the pilgrims. Those who come from Chhindwara side pass through Imlidhana, Alimodh and Rorighat to reach Kajri on foot. Pilgrims from Maharashtra and other parts of Madhya Pradesh come to Pachmarhi town by car/bus. They come to Dhoopgarh by local jeep and then start journey on foot via Kalajhar, Nagphani to Kajri.

Apart from the religious tourism, agriculture is the other main occupation of the villagers. Majority of the agricultural fields are irrigated. They channelise stream water through plastic pipes to irrigate their fields. Rice, wheat, maize, kodu, mustard seed, etc., are their main crops. They also grow eggplant, different types of

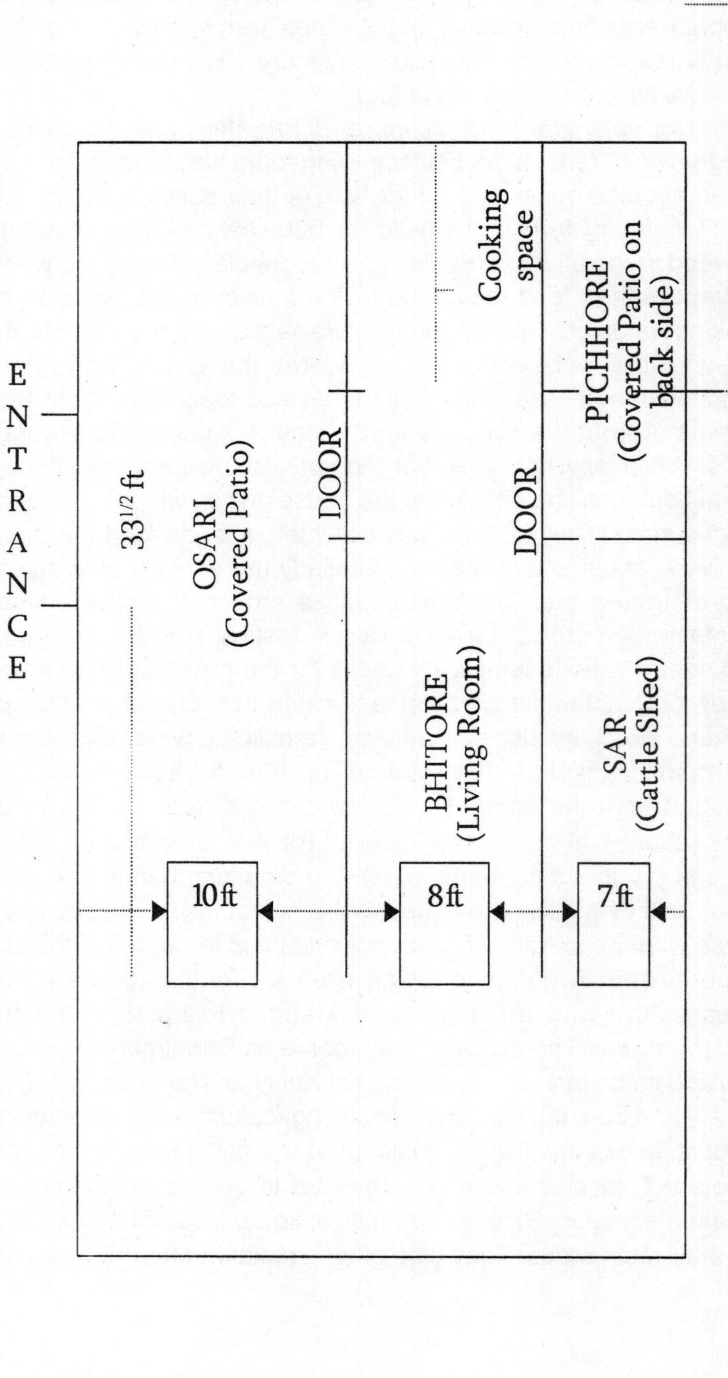
ENTRANCE
$33^{1/2}$ ft
OSARI
(Covered Patio)
DOOR
BHITORE
(Living Room)
Cooking
space
DOOR
PICHHORE
(Covered Patio on
back side)
SAR
(Cattle Shed)
16 ft
10ft
8ft
7ft

Fig : 1

bean and leafy vegetable in their kitchen garden. They also supplement their income through the selling of non-timber forest produces, cattle rearing and ghee (clarified butter). Forest department has constructed several water storage tanks, which receive stream water through pipes. Villagers collect drinking water from these tanks. They also use water from tube wells for the same purpose. Before 1977, Kajri had 1005.754 hectares of land. At present it has 14.629 hectares of land, i.e. the villagers have lost 98.54% of their community land to the forest department and Satpura Tiger Reserve Project without any compensation.

Binora Village :

Binora is an uniethnic revenue village, inhabited by the Mawasi. It is situated in the southern direction of Pachmarhi town. There is a forest route from Rorighat village to Binora. Road condition is so bad that even four-wheel driven petrol jeep cannot go there for crossing a hill. Only the tribal people use this foot track during dry season. Survey team had to travel 106 km to reach this village in a long all weather route (i.e. from Pachmarhi Matkuli Delakhari Kuabadla Kursidhana Bhurabha Binora. From Bhurabhagat onward road is not metalled and before entering Binora village one has to cross the Denwa River either on foot or car. Like other villages, stated earlier, this village had 764.202 hectares of land before 1977. At present it has 61.697 hectares of land, i.e. 91.92% of their community property has been taken away by the forest department without giving any compensation. The Denwa River forms the western and southern boundary of the village. On the north, there is a hill and the entire area is a part of the Satpura National Park. In the east is Nadia village, which comes under the buffer zone.

Binora village has 41 families. Average family size is 5.8. The main village street runs from east to west.

All the houses are mud built and face towards the village street and exhibit scattered cluster type settlement pattern. Each household has a cattle shed on its side. Agriculture, due to soil type is very poor (i.e. sandy and full of gravels), is the primary occupation of the villagers. They use river water for irrigation. River lift pumps are also used in some cases. They grow wheat, maize, kodu and kutki in their fields. Even some of them practice *dhaiya* (shifting cultivation) by clearing lantana bush in the forest. As they clear lantana bush only, the forest department does not object much because it has to spend a lot annually for the same purpose. But the only difficulty in this process is that the topsoil often gets eroded every rainy season.

Besides agriculture, they sell ghee and livestock etc., in weekly market at Bhurabhagat to supplement their income. They collect firewood from the forest for domestic consumption only. They also collect other forest produce (though there is a ban on some items like tendu and mohua leaves, medicinal plants, edible roots etc.,) and sell them to the representative of the agents at Bhurabhagat. Making of country liquor is another important economic activity of Binora village. They collect mohua flower from forest. Dried mohua flower are kept in big vessels like iron barrels or in big aluminum vessels along with water. After two/three days, fermentation takes place. They take out the fermented liquid and distill it to make country liquor with their indigenous technology on the bank of the Denwa River. Mohua liquor has a steady demand in the surrounding areas. Moreover, during Shivratri and Nagpanchami festival, pilgrims from Chhindwara district (Madhya Pradesh) and Maharashtra state often pass through this area. They mainly come on foot. They consume a lot of this stuff. One distillation plant requires about 10 kg. firewood per day. On a normal day, 4/5 distillation plants operate together on the bank of River Denwa. It was reported that the villagers often buy dried mohua flower from the market, as their own collection is not sufficient from the point of view of production and demand in the neighbouring areas.

This village has no electricity connection. It has a single teacher primary school.

Pattan Village:

Pattan is a revenue village situated on the bank of Denwa River. It is about 126 km away from Pachmarhi town in the western direction. Due to presence of the Satpura Hill range between Pachmarhi and Pattan, one has to travel to Piparia, then to

Map - 7

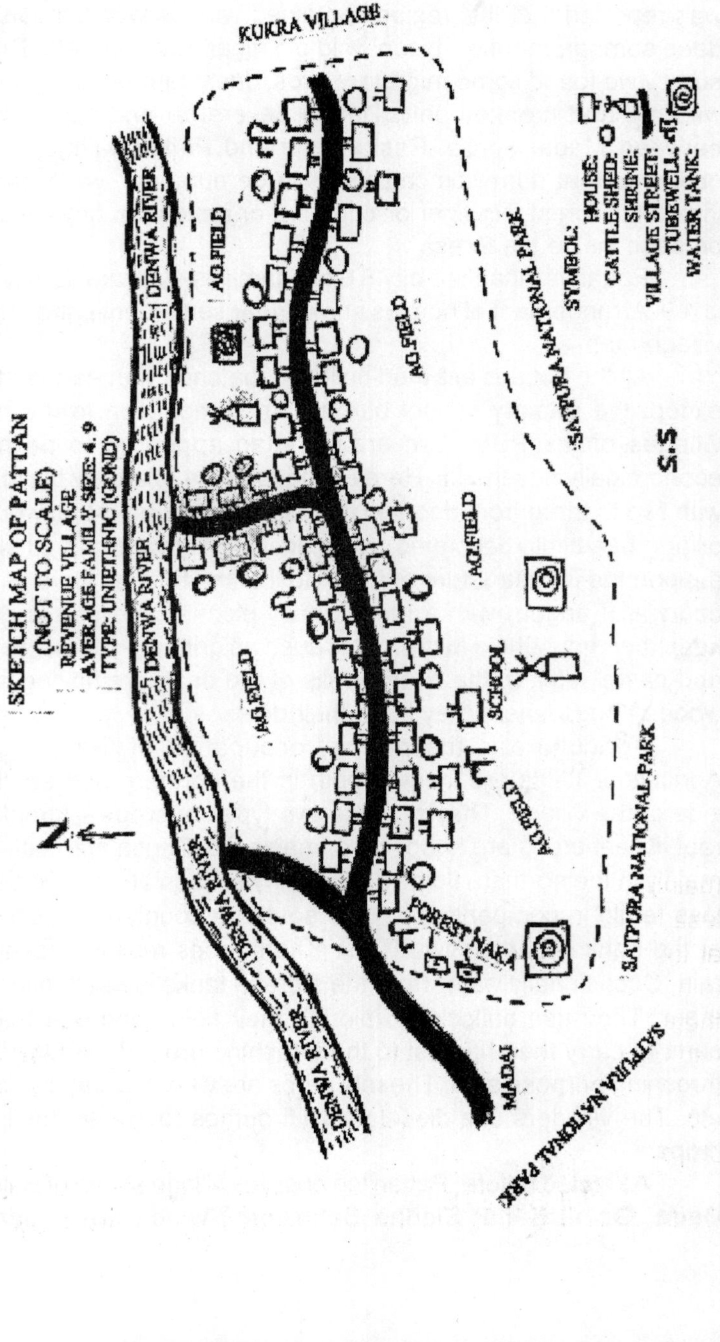
SKETCH MAP OF PATTAN
(NOT TO SCALE)
REVENUE VILLAGE
AVERAGE FAMILY SIZE: 4.9
TYPE: UNIETHNIC (GOND)
N
KUKRA VILLAGE
DENWA RIVER
AG.FIELD
SCHOOL
FOREST NAKA
MADAI
SATPURA NATIONAL PARK
SYMBOL:
HOUSE:
CATTLE SHED:
SHRINE:
VILLAGE STREET:
TUBEWELL:
WATER TANK:
S.S

Sohagpur and Madai. From Madai it is about half an hour journey on the unmetaled narrow road through relatively dense forest. It was reported that this region contained various wild animals like deer, sambar, monkey, bison, wild pig, tiger, wild dog, etc. During survey we found some migratory birds, deer, sambar, a few bison, wild pig and monkey only. During several to and fro journeys between Madai Forest Rest House and Pattan village, it was observed that domestic cattle's, in large numbers, were grazing inside the forest. No tiger or pugmark or residue of any kill could be seen inside the forest.

Pattan is inhabited by 48 Gond families. Average family size is 4.9. Arrangement of houses show linear settlement pattern (vide village map.

All the houses are mud built and thatched with earthen tiles, except the primary school building. In comparison to the other villages of the PBR core area, Pattan appeared to be more economically advanced. Here the houses are relatively big in size with two to three front doors in the *osari*. Osaris are stout wooden pillars, beautifully decorated with chemical paints and carved which support the infrastructure of roof. Most of the houses have fenced courtyard (angan) with a sacred basil plant at one corner. They keep their household instruments such as grinding stones, mortar and pastel, etc., in the osari. Walls of the osari are having small wooden pegs where they hang their dresses.

Agriculture is the primary occupation of the villagers. Agriculture fields are located both in the northern and southern side of the village. They produce two types of crops – kharif and rabi. Kharif crops are kodo, kutki, maize etc. These are cultivated mainly in the northern fields during rainy season and the fields are less fertile in comparison to their southern counterparts situated at the bank of the Denwa River. Kharif crops mainly depend on rain. Occasionally water from the storage tanks is used to irrigate them. They use bullocks to plough their fields and use bullock carts to carry their harvest to the threshing ground, and lastly for threshing purpose also. The rabi crops are wheat, gram, mustard, etc. The villagers use diesel river lift pumps to irrigate their rabi crops.

As stated before, Pattan too possess village shrine of Bajrang Dada, Goyal Baba, Siddha Baba, etc., which are important landmarks in a Gond village. They also follow the same marriage rules related to une and pure clans strictly. Hypergamy is tolerated whereas hypogamy is strictly prohibited. Pattan village has no electricity. Here also a single teacher looks after the primary school students from class-I to class-IV. As reported, students get midday meals regularly.

Jambudweep Village:

Jambudweep is a revenue village. It is situated about 7 km away from Pachmarhi town in the northern direction. From the forest gate of Satpura National Park there is a metaled road upto the 'Twynam Pool' about 1 km. Twynam Pool was inaugurated by His Excellency, Dr. Henry J. Twynam, Governor of Central Provinces and Sera on May 07, 1945. It is a standard swimming pool and gets its water supply from the hill stream. During winter season, this pool remains dry. It is a famous tourist spot located inside the PBR core area. During summer month's, tourists are allowed to go inside and take a cold bath in the pool by paying a nominal fee. Beyond Twynam Pool, there is no road. On the undulated rocky terrain one has to go on foot a little over 2 km to reach Jambudweep village after crossing a beautiful perennial hill stream.

According to the forest officials, Jambudweep is an uninhabited (viran) village and the same is recorded in their documents. Before 1977, Jambudweep had 180.901 hectares of land. It has 1.986 hectare of land at present. It shows that the Madhya Pradesh Forest Department has taken away about 98.90% of land from Jambudweep. In the beginning it was a puzzle for us that if the said village has no habitation inside the core area, forest department could take the whole area. Why they left out a little over one hectare of land there? While conducting our field investigation in other revenue villages, we asked the villagers about Jambudweep. We were told that Jambudweep has one Yadav family. They still live there. So, our survey team decided to visit the area. Journey to Jambudweep is very difficult because one has to negotiate steep rocky slopes at several places.

Jambudweep is a beautiful bowl shaped valley. It is inhabited by a single Yadav family. Right before independence, a

British administrator was in-charge of Pachmarhi area. He had an Indian valet (khansama) named Mr. Danish. Local people refer him as Danish Sahib. He was unmarried. He used to accompany his master during hunting. He visited Jambudweep area many times. During British period, a small check gate was constructed on the Jambudweep stream to stop water so that the washer men of Pachmarhi could wash clothes there. Mr. Danish liked the place and wanted to settle there after his retirement. So through his master's help he was able to register some land in his name. He used to pay revenue taxes for the same. After his retirement, Mr. Danish took up farming at Jambudweep. Though his fields were well irrigated, production was relatively low, due to sandy soil. So he invited a milkman (Yadav) to come and settle on his land with his cattle without any rent. It was agreed upon that the milkman would provide cow dung to his fields and help him in his agricultural operations. It continued for many years and gradually Mr. Danish (in his old days) became fully dependent on the Yadav family. The youngest son (the present occupant) of the milkman was very fond of Mr. Danish and used to address him as 'Dada' (grandfather). Before his death, Mr. Danish transferred all his farms in the name of the present occupant who was a little boy. Since then, the present occupant is living there and paying his taxes and his name is duly recorded in the register of the State Settlement Department. With the creation of Satpura National Park, this area came into core area of PBR. Forest department first dismantled the check gate where the washer men used to come for washing clothes. Then they ordered the Yadav family not to graze their cattle in the forest. Lastly they asked the Yadav family to vacate the land. As the person is having his patta (ownership record of the land) so the forest department could not evict him. They live in a very dilapidated hut, which may fall at any time. As due to court order, no new construction can be made inside the PBR core area, so the forest department is not even allowing them to repair their old hut.

In half of their land, they have planted mango and guava trees. They have also planted a good number of lemon plants. In the rest of the field they produce various types of vegetables like eggplant, spinach, mint, coriander, gourd, pumpkin, chilly etc. Besides, they sell milk and ghee (clarified butter) in Pachmarhi market.

Village wise primary as well as supplementary economic activities in the PBR core area are given below :

Sl. No.	*Village*	*Primary Economic Activities*	*Supplementary Economic Activities*
1.	Bariaam	(i) Collection of firewood from forest and selling it in Pachmarhi market.	(i) Daily wage labour (clearing lantana bush in forest, road repairing, working as helpers in the shops and hotels of Pachmarhi town during festivals. (ii) Cattle rearing and selling of ghee (clarified butter) (iii) Renting out patios to pilgrims during Maha Shivratri festival. (iv) Setting up of roadside shops during Maha Shivratri festival.
2.	Pagara Dhana	(i) Cultivation	(i) Collection and selling of fire wood. (ii) Daily wage labour (clearing lantana bush, cleaner in truck, agricultural labour in plain areas during harvesting season). (iii) Cattle rearing and selling of ghee (clarified butter) (iv) Collection and selling of NTFP. (v) Selling of fruits (guava, mango, banana, etc.) from Orchard.
3.	Ghoranar	(i) Cultivation	(i) Same as stated above in the case of Pagara Dhana village, except selling of fruits.
4.	Badkachar	(i) Cultivation	(i) Same as Ghoranar village.
5.	Rorighat	(i) Cultivation	(i) Daily wage labour. (ii) Collection of NTFP. (iii) Cattle rearing and selling of ghee (clarified butter) (iv) Setting up of roadside stalls during fairs and festivals.
6.	Kajri	(i) Renting out rooms to the pilgrims during Nag Panchami festival.	(i) Cultivation. (ii) Collection of firewood for domestic use and for selling to the pilgrims. (iii) Daily wage labour (work as porter during Nagpanchmi). (iv) Collection and selling of NTFP. (v) Cattle rearing.

(Contd.)

(Contd.)

Sl. No.	Village	Primary Economic Activities	Supplementary Economic Activities
7.	Binora	(i) Cultivation	(i) Preparation and selling of country liquor from Mohua flower to the pilgrims during fairs and festivals. (ii) Collection and selling of NTFP (iii) Daily wage labour. (iv) Cattle rearing.
8.	Pattan	(i) Cultivation	(i) Cattle rearing. (ii) Daily wage labour. (iii) Collection of firewood from forest.
9.	Jambudweep	(i) Cultivation	(i) Cattle rearing and selling of ghee (clarified butter). (ii) Selling of fruits from orchard. (iii) Collection of NTFP.

People's Problem

In this chapter attempt has been made to highlight different problems of the villagers living within the PBR core area. As the study team had to cover nine villages within a relatively short span of time so it was decided that in each village, people's view would be collected through group discussion regarding their day-to-day life activities. These have been supplemented through our observations and other secondary analysis. This chapter has three sections as given below.

(a) People's view
(b) Nature of anthropogenic pressure on the forest
(c) Condition of tourism in Pachmarhi

(a) People's view

Life in the PBR core area is really hard. Tribals have no social stratification or hierarchy among them even though some economic differences prevail there. Outsiders can easily identity from the house types about the economic condition of their owners. Relatively economically well off families have good huts thatched with earthen tiles (khapra) and on the other hand the poor people live in mud huts thatched with dried grass or straw. In majority of the cases, the later types of huts are in bad shape. It was reported that following a Madhya Pradesh High Courts order, there is a blanket ban on any new construction or any repair or renovation of old structures in the entire PBR core area along with Pachmarhi town.

Case Study I: Village – Jambudweep

In Jambudweep, the house of Shri B. Yadav (40 years, male) is in a state of dilapidation. Due to forest departments' restriction, he cannot construct a new house on his own land for which he pays tax (lagaan). With a sad smile on his face he stated that if this situation continues for some more years, his family might get buried under the roof of his own house. He also informed that the Satpura Tiger Reserve Authority has already shown this revenue village as 'uninhabited' (viran) in their office records. Forest department has already taken away almost all the village land. If his family dies then the department could get 100% of this revenue village under their control without giving any compensation. Shri Yadav has a physically handicapped son and a mentally retarded daughter in his family. He is always worried that, in his absence what would happen to his children.

He has a small plot of land. Soil type is sandy with low water retention power and reddish in colour. However, he manages to irrigate his field continuously with the help of indigenous technique by diverting the flow of a natural perennial stream. He has never received any assistance from the government administration in the form of seed, fertilizer, pesticide etc. He uses cow dung as manure. It was observed that half of his land is utilized for growing seasonal vegetables and cereals and in the other half he has planted twenty-one mango, guava and lemon trees. With these fruit bearing trees and bushes his homestead looks greener in comparison to the adjoining areas under forest department's control. The forest department always exerts pressure on them to vacate the place. Shri Yadav told that he is always ready to vacate this area if he is given suitable rehabilitation elsewhere, but forest department is not taking any positive steps in that regard. He further expressed that his life is full of difficulties, as Jambudweep village has no road, safe drinking water, electricity and healthcare facilities. Even to sell his agricultural products, he has to walk down seven kilometers to reach market at Pachmarhi town.

Case Study 2: Village – Bariaam

In Bariaam village, during group discussion, a group of old villagers were interviewed. In their opinion, their village has become very small at present than what they had seen in their childhood days. In the past, they could easily collect necessary forest produces and firewood nearer to their village. On the northern side of the village now exists the new State Public Works Department (PWD) road, locally known as Pachmarhi-Pipariya Road. During formation and notification of the Pachmarhi Sanctuary and the Satpura National Park, most of their village land has been taken away by the forest department along with the village lake. Now they have no fishing right in the Bariaam Lake. Similarly all the agricultural and Dhaiya (shifting cultivation) plots have been acquired by the forest department. As a consequence, out of 22 families, only three families have cultivable plots adjacent to their huts. Shri B. Singh (50 years, Male) of Bariaam said, ***"Pehle gaon bahut bada tha. Jangal mein lakdi, ghaas, baans bahut tha. Kheti hota tha. Ab kuch nahi hain. Pachmarhi sahar ke najdik gaon hone se labh hain. Shiksha, swasthya seva, bus, taxi ane jane ke liye milta hai. Gaon ke beech se purana Pachmarhi Pipariya sadak tha. Ab naya sadak ban gaya hai. Pehle jangal mein jane tatha kam karne, janwaar charane, lakdi le aane mein koi rok nahi tha. Sabhilog kheti karte the, sabji, bhaji, lagate the. Doodh bhi bahut hota tha. Jabse Pachmarhi Sanctuary bana tabse gaon ka jamin forest deparment le liya. Kheti ka jamin khatam ho gaya. Jangal se lakdi lena, ghas lena, janwar charana mushkil ho gaya. Pade likhe ladke ladki***

ko naukri bhi nahi mila. Pachmarhi mein bhi bheer ho gaya. Lakdi ke mang bad gayee, lekin jangal mein lakdi kam ho gaya. Din bhar ghoomne se 20-25 kg lakdi milta hai jise doosre din bech kar Rs. 25-30/- milta hai. Usi se samaan kharid kar le aate hain. Sabji, dudh pehle jaisa nahi hota. Jeevan me taklif ho gaya hai. Sansaar chalane mein bania logo se udhaar lena padta hai. Taklif dekhnewala koi nahi hain".

"The area of our village was large before. There were plenty of firewood, fodder and bamboo. We used to do cultivation. We get some benefit as our village is situated near Pachmarhi town. We get education, healthcare, transportation facilities in the form of bus and cab. Old Pachmarhi Pipariya road was running through the middle of our village. Now new road has been constructed (outside our village). In the past there was no restriction to go inside village forest for doing work, grazing domestic cattle and fetching firewood. Everybody used to cultivate their fields and also were growing vegetables, leafy vegetables etc., in their kitchen gardens. Since the declaration of Pachmarhi Sanctuary all the agricultural fields have been taken away by the forest department. Rules regarding collection of firewood, fodder etc., and grazing of domestic cattle have been changed. Educated boys and girls are not getting any employment. Pachmarhi town has become more crowded. Demand of firewood has escalated but firewood has become less. One can collect 20-25 kg firewood from forest after a whole days search. Next day they are sold @ Rs. 25/- to Rs. 30/-. We buy necessary things with that money. We cannot produce milk, vegetables, etc., as we used to do in the past. Life has become difficult for us. Sometimes we get chance to work as day labour. Often we have to take loan from shopkeepers to run our families. There is nobody to help us".

The other villagers who were present in the group discussion supported Shri B. Singh's views and nodded in unison. They further reported that forest is necessary for their survival. For the sake of joint forest management (JFM) and for the development of their village they have Eco Vikash Samity (Eco Development Society) in their village. This society has representatives of both forest department and villagers in its governing body. Unfortunately the said society is totally defunct now and doing nothing for the welfare of the villagers. Deforested areas have been left alone for natural regeneration. Rate of natural regeneration is very slow. Layer of topsoil is thin. In many parts it has been eroded by rainwater. Roots of plants, irrespective of their ages, got exposed. For collecting firewood, the villagers have to move around 6-8 km on an average. Apart from daily wage earning works, collection and selling of firewood is the major occupation at Bariaam village. They consider themselves lucky for being nearer to an urban settlement where they get occasional employment. Otherwise their life would have been much more miserable.

They reported that they have been better literary rate in the village as they can send their children to Pachmarhi town or to other towns in Madhya Pradesh. However, in the question of rehabilitation there were mixed responses. Relatively economically well off families do not want to leave the area and are more or les satisfied with the present condition. Informants belonging to the poorer families stated that they have no objection at all to settle elsewhere if they are given proper compensation and facilities to earn their livelihood.

Case Study 3: Village – Pagara Dhana

In Pagara Dhana village, during group discussion the villagers reported about the following difficulties.

(a) They cannot repair their houses.

(b) Do not get sufficient fodder for their cattle.

(c) Collection of forest produce has become difficult partly due to reduction of forest resources around them and partly due to restriction imposed upon them by the forest department.

(d) No medical facilities. Sick persons have to be taken to distant places for medical treatment.

(e) Safe drinking water is not available.

The oldest person the village, Shri D. Singh (75 years, Male) said, "Hamere jeevan mein bahut taklif hai. Makaan ka halat to dekh rahe hain. Maramat bhi nahi kar sakte hain. Janwaron ko charane ke liey jangal mein nahi le ja sakte hai. Fores dipat ka rok hai. Barsha mein ghar mein pani chuta hai. Jam ham apne bachpan ka din yaad karte hai to bahut dukh hota hai. Us samay ghar tatha sar ka chat mota mota ghas dalkar banta tha. Garmi nahi lagta tha. Barsa mein taklif nahi hota tha. Janwar ke saath jangal mein din bhar rehene se bhookh nahi lagta thha. Kyonki jangal mein bahut fal hota tha. Ab utna nahi milta. Jangal badal gaya. Ab log plastic laga kar patla ghas chat par dalte hain. Chat khapra kharid kar bhi bana sakta hai. Lekin iska daam adhik hai. Bachche logon ko naukri nahi hai. Raste ka haal aap dekh hi rahe hain. Barsaat mein chalne layek nahi hai. Nala mein pani bhar jata hai. Gai, bakri lekar bahar jana kathin hai. Achcha to yeh hai ki jangal se idhar bhalu, cheeta, sher kabhi nahi aata. Beemar padne se dawa daru ham khud hi kar lete hai. Jyada kharab hone se aspatal me le jana padta hai. Gai, bakri ka chara bahut kam ho gaya hai. Ab uske liye bahut duur tak jangal ke andar ghusna padta hai. Jangal mein pehle bans adhik tha jisse bahut cheej banta tha. Ab bans bahut kam ho gaya. Pathar ka tukra jo ghat banane ke liye lagta hai usme rok tham nahi hai. Lekin unko jodne ke liye chikni mitti bahut duur se lana padta hai. Kheti ke upaj bahut kam hai. Kheti ke lal mitti retila hai, keval chota chota pathar bhara hai. Gaon mein ischool nahi hai. Bachche logo ko chota ischool se bada ischool jane ke liye koi subidha nahi hai. Dhaiya karna bilkul band hai. Thelua jamin mein amrud, nimbu, aam ka ped lagaya gaya hai. Usi mein murgi bhi koi koi palta hai. Isse kuch kuch aamdani hota hai. Jangle mein koi kaam nahi hai. Bachche logon ka chahat badta hai, aamdani nahi. Mahangai bhi badta ja raha hai. Naye bachche Pachmarhi Pipariya ja kar naya naya samaan dekte hain, usko pane ki ichcha rakhte hain. Shadi ristedari mein bhi karcha bahut ho raha hai. Kam chalane ke liye baniya samaj se udhar lena padta hai. Jiwan kathin hota ja raha hai".

"There are lots of difficulties in our life. You can very well see the state of our house. We cannot even undertake repairs. We cannot take our animals to graze in the forest. Forest department have imposed restriction on that. The roofs leak in the rain. When we remember our childhood days, we feel very much sad. In those days our houses and the roofs were made by placing thick layers of grass. We did not feel the heat or the difficulties of the rain. We used to graze our animals in the forest day long without feeling hungry because there were lots of fruits in the jungle. Nowadays it's very limited. The forests have changed. The people use plastic and thin layers of grass to make the roof. Earthen tiles can also be bought for the same but it's costly. Our children don't get jobs. You can see the condition of our roads. It's impossible to walk in the rainy season. The streams are full with water. It's difficult to go out and graze the cows and goats. It's good that the bear, leopard or tiger do not come here. In case of illness, we usually treat ourselves. However if it worsens, we go to the hospital. Fodder of cows and goats has diminished. We are compelled to go deep into the forest for it. There were lots of bamboos in the forest from which many articles are made. Now the bamboos are very less. There is no restriction on stone pieces used for construction of houses but the clay required for filling has to be brought from distant places. Agricultural yield is very less. The red soil in the fields is sandy which are full of small pebbles. There is no school in the village. There is no facility for the children to go from primary school to secondary school. Shifting cultivation has been totally banned. In the vacant land, guava, lemon and mango have been planted. Some rear hen, which is a source of income. There is no work in the forest. The demands of the children are increasing but there is no source of income. Market price is increasing gradually. The young people who go to Pachmarhi Pipariya see new things and desire to have it for themselves. There are lots of expenditure on marriage and social obligations among relatives. To meet up with expenditure, we have to take loans from the businessman caste. Life is gradually becoming tough".

The villagers opined that forest is very necessary for their survival and should be protected. They have no objection to leave their native village but they want to get rehabilitated in any suitable place with proper compensation. Forest department has informed them about this but have not taken any concrete steps for the core area villages. They sincerely wish that the forest department would undertake early initiative so that they may not be subjected to any further hardships.

Case Study 4: Village – Ghoranar

There is a general ban on the collection of tendu leaf, mahul leaf, safed musli, different medicinal plants and herbs, long grasses for thatching roof and preparing brooms, honey, etc., from the forest. Apart from daily wage labour, rearing of cattle and very insignificant cultivation of crops, collection and selling of the above mentioned forest produces help the people of Ghoranar village to keep their body and soul together with great difficulty. During survey, the investigators wanted to get more data on these in a group discussion. The villagers reacted sharply and declined to say anything. Shri V. Singh (45 years, Male) as their spokesperson said, "We are already having a very tough time. We don't want to invite any further trouble for us by sharing this information with you. This helps us to live. We will die out of starvation if your reporting stops this. We have already told you a lot of things in good faith. Don't ask any details. Please excuse us".

Ghoranar is a forest village, which was established by the forest department for extraction of timber. After 1977, cutting of forest for timber has been totally stopped. In the group discussion, Shri B. Singh (35 years, Male) said, "Jis jangal ko bachane tatha jangal se desh ke vikas me lakdi nikalne ke liye hamare dada, babuji, bhai log apna jiwan bitaya, aaj us jangal ka kam bandh ho jane se hum logo ka jiwan kathinayee me ho gaya hai. Woh log jangal ke bhitar ja ja kar achcha achcha ped dekh kar katke le aate the tatha jangal ko hara bhara karne me hamesha lage rehte the. Usi jangal me ab apna manse ghumne, kam karne tatha jiwan chalane me rok lag gaya hai. Naya cheej dekha nahi tha. Socha tha hawa saaf hai, kam jeene ke liye milta hai, kahi aur dekh ke kya karenge. Yehi achcha hai. Lekin park banane tatha jangal bachane ke nam par hame bahut kast ho raha hai. Bal bachcha kaise bachege, chinta hota hai".

"The forest to save which and from which for national development our grandfathers, fathers brothers extracted timber and spent their life, today have brought hardships in our life because of the restriction in work over there. They used to go deep inside the forest and selectively cut trees and bring and at the same time engaged in keeping the forest green. In the same forest today there is restriction on roaming, working and earning a livelihood. We hadn't thought of seeing anything else. We thought that with the fresh air and work available for survival, why to go elsewhere searching. This is good enough. But with the coming of the park and in the name of saving forest, we are facing a lot of trouble. We are worried, how will our children survive".

Case Study 5: Village – Badkachar

Like Ghoranar, the villagers of Badkachar also expressed the same opinion in the course of group discussion. Shri J.P. Singh (50 years, Male) stated, "Jangal mein ab kya rakha hai? Har kadam mein takleef he takleef hai. Rojgar ka sadhn kuch nahi hai. Majduri ka kam bhi nahi milta. Logon ka jeena ab bahut mushkil ho gaya. Is halat mein apko kya bataye. Mere vichar mein jangal chodne se hi jinda rah payenge".

"What is now left in the forest? There are hardships everywhere. We have no source of income. We do not get work as wage labourer. It has become difficult to make our living. What can we further say in this condition? In my opinion, we can only survive if we leave this forest".

In the same group discussion, Shri B. Singh (40 years, Male) said, "Gaon ke chhetra pahele se bahut chhota ho gaya hai. Kheti bari me bahut takleef hota hai. Kheti bari bhi kam ho gaya hai. Abhi kheti jangal ke chhetra mein kuch bacha hai. Koi koi udhar ja ke kheti karne ka prayash karte hai to forest dipat ke taraf se unko bahut kuch bhugatna padta hai. Aaj ke samay mein bhi hum logo ko kutki aur kodo kha kar jeena pad raha hai. Kahi sambandhi se milne tatha jarurat pad ek ek din jangal ke raste se paidal chalna padta hai. Jangal mein ab mangal nahi hai. Behtar to yehi hoga ki kuch achcha jagah mile aur bal bachcha lekar ek naya koshish shuru kiye jaye. Suna to hai ki sarkar is bare mein soch rahe hai. Lekin abtak pura jankari nahi mila hai".

"The area of the village has reduced from before. There are lots of difficulties in cultivation and the yield is very less. Only a few cultivable plots are left in the jungle. Some people go there and try to cultivate. But they have to face a lot from the forest department. Even now we feed on kutki and kodo to survive. We have to walk a whole day through the forest to meet our relatives and in time of necessity. Nowadays, the forest does not cater to our welfare. It would be better for us if some good place were allotted where we can begin our new life along with our children. We have heard that the government is thinking in regard. But till now we do not have the complete information".

Case Study 6: Village – Binora
Informant: Shri C. Singh (45 years, Male)

"Our village is located in a corner of the Satpura National Park. It is a border village between Hoshangabad and Chhindwara districts. The Denwa river separates our village from Chhindwara. ... Out of 42 households only five families have some cultivable land in our village. Soil condition is not good. It is sandy and full of small stones (gravel). From Nadiya (which is about 12 km away from Binora) to our village, density of forest is very thin. There is no good road and conveyance. Every Tuesday all of us have to walk 5 km to attend weekly market at Bhurabhagat. Residents' of other villages, such as Alimoth, Bagoli, Dholam, Rorighat, Dobansarra, Kuchikoh, Pipar Pathar, etc., come to the weekly market. We have to cross Denwa River on foot to reach Bhurabhagat. It becomes very difficult for us during rainy season. For going out to other places like Matkuli, Pipariya, Bhopal, Hoshangabad etc., we have to go to Nadiya on foot to avail private or state owned transport facilities. We have a primary school in our village. Our children cannot continue their studies beyond primary level because we cannot send them to distant places for secondary education. It requires money, which we do not have.

In our village two families have diesel pump sets to irrigate their agricultural fields. They use river water. For purchasing fuel for their pump sets they have to travel about 20 km either to reach Tamia or Ranikheda. The other villagers are mostly daily wage labour. About two third families of our village prepare country liquor through distillation of fermented mohua flower juice. In the weekly market they sell mohua liquor. During Shivratri festival about hundred thousand pilgrims pass through Bhurabhagat area on foot to reach the Chouragarh Hills at Pachmarhi. At Bhurabhagat a fair is organized for two weeks during that time. To supply country liquor to the pilgrims our villagers work day and night for one month.

The pilgrims usually camp on the banks of river Denwa. They prepare food, take bath, wash their dresses, stay overnight on the river banks in make shift camps and use riverbanks for sanitation.

Particularly during this period the river water gets polluted and often diarrohea breaks out. We avoid drinking river water during the festivals. At the end of the festival we collectively worship Mailee mata inside the forest so that nobody falls ill in the village."

Shri M. Singh (40 years, Male) of the same village said that, "Kheti ka jamin nahi hai. Majduri bhi nahi milta hai. Jo kuchh jangal mein hai usme se kanda, lakdi aur janwaro ka chara pura nahi hota. Jab kabhi gaonwale nadi par karke Chhindwara ke jangal mein jate hai to forest ke log takleef dete hain. Gaon ke paas jangal mein mahua utna nahi hai. Is liye bazaar se mahua khareed kar daru banana padta hai. Gaon mein pahele doodh hota tha. Ab to charo taraf khali pathar hai. Yahaan jiwan bahut kathin hai. Gaon chodke jana hai to chale jayenge. Agar sarkar koi achcha jagah mein basaye aur kuch subidha de to bal bachcha sab ke liya theek hoga".

"There are no agricultural lands. We don't get daily wage work. Edible roots, firewood and fodder for cattle available in the forest do not suffice our needs. Whenever we try to cross over the river to the Chhindwara forest, the forest officials harass us. The mohua trees are scarce in the forest near our village. Therefore we have to buy mohua leaves from the market to prepare liquor. In the past there was milk in the village. Nowadays there are stones all around. Life has become difficult here. We will leave the village if we have to. It would be good for all the children if the government rehabilitates us in some better place with good facilities".

Case Study 7: Village – Kajri

Informant: (1) Shri R. Singh (27 years, Male)
(2) Shri K. Singh (50 years, Male)

"Hamara gaon park ke bilkul ek kinare mein hai tatha charo taraf se paharo se ghira hai. Kheti ka jamin pahele bahut tha. Jangal khula tha. Falwale ped, madhu, chara jamin bahut tha. Nagdwari mandir par Nag baba ke puja hetu Nagpur se bahut log aate hai. Sabhi ko mere gaon me rahana padta hai. British jamane ke pahele se suna hai log is puja ke liye paidal Pachmarhi hokar aate hai. Ab bhi chale aa rahe hai. Doosra rasta nahi hai. Gaon mein saal bhar ke khane ke liye anaj nahi hota hai. Kheti ke saath hum long ghee banate hai, jangal se madhu le aate hai, rambahuri ghas bhi le aate hai. Is ghas ke jhadu bante hai. Yeh cheej bazaar mein bech kar hum log ka kam chalta hai. Nagpanchami mein aanewale logon ko bhi humlog yeh samaan bechte hai. Hum jangal se lakdi bhi le aate hai. Apna khana banane, sardi mein ghar garam rakhne ke liye tatha Nagpanchami mein aanewale yatriyo ko khana banane ke liye lakdi kam aate hai. Yatriyo ko ek chhota bandil lakdi paanch rupaye main bechte hain. Yeh log hamare ghar mein rahte hai. Har aadmi se har roj dus rupiay lete hai. Is gaon ke sabhi ghar mein lamba lamba osari hai. Ek ek osari mein 30-40 aadmi sote hai. Yeh kam kewal do hafte tak chalta hai. Madhu ka ek botal bechne se sau rupiya milta hai. Madhu hum log pahar se late hai. Yeh bahut takleef ka kam hai, lekin jeene ke liye takleef uthana padta hai. Isme ek samuh kam karta hai. Ek samuh mein dus aadmi hote hai. Madhu lane ke liye do din lagta hai. Samuh ke har aadmi ko barabar hissa milta hai. Ek aadmi ko kam se kam char tina milta hai. Forest se lantana katne tatha aag rok ne ke liye line lanane kabhi kabhi kam milta hai. Ek din ke pachash rupiay majduri milta hai. Nagpur ke mandal ke log aur tirath yatri ke samaan dhone ke liye bhi coolie ka kam milta hai. Mela mein kamane ka aur bhi rasta hai. Gaon ke log chai, paan aur bidi, tambaku ka dukan, puja ke samaan ka dukan lagate hain. Pahar chadne mein asamarth tirath yatri logo aur unke samaan ko dho kar Nagdwari Pahar ke upar le jane se majduri milta hai. Nishangarh pahar par yatri log nishan lagate hai jinke saath rasta batane mein majduir milta hai. Mahadev Mela Samity aur forest deparment milkar tirath yatrion ke liey dam banaye hai. Isse hum sab ka achcha hua. Peene ka pani aur khet mein sichai ka pani ab bahut milta hai. Park banne ke bad jangal se samaan le aane mein rok lag gaya hai aur kheti ke jamin kam ho gaye hai. Naya ghar banane tatha purana ghar maramat karne par bhi rok lag gaya hai. Agar nagdwari jaisa tirath sthal nahi hota to hum log ab tak khatam ho gaye hote. Gaon mein kya hai? Kuch nahi. School nahi hai, rasta nahi hai, ilaaj ke liye koi subidha nahi hai. Dawa to hum khud karte hai. Jangal jadi-buti ka prayog karte hai. Jab kabhi Pachmarhi, Bhopal tatha Nagpur jate hai to waha ka jiwan aur subidha dekh kar apne gaon mein taklif hota hai. Sahar mein rahene ke liye padhai tatha buddhi ki jarurat hai. Naukri hone se bhi subidha milta hai jo ki hum logon ke pass nahi hai. Sarkar tatha forest department chahe to achcha sadak, aspatal aur school khol de, jangal se madhu, achar, jadu ghas lene ka chuut de, saath mein Nagdwari, Chitrasla, Nishangarh, Kalajhar ko paryatak sthal bana kar subidha de to apna gaon sabse achcha ho jayega. Jo kam hum log jangal mein karte hai tatha haath se dus hazaar rupiya har sal kamate hai isse bhi achcha kamayenge tatha jangal ka bhi dekhbhal karenge. Saath mein purane samay se chala aa raha mela bhi achchi tarah se chalegi, Suddh hawa, suddh pani, pahari rahan sahan, dharmik sthal milkar gaon achcha paryatak sthal ban jayega. Isme sabho ka bhala hai. Ab to jo ho raha hai woh theek nahi hai. Jangal mein rahenewale logon ko jangal jane se rok lagne se jangal se duree badti ja rahi hai".

"Our village is situated at one end of the park and surrounded by the mountains on the four sides. Agricultural land was plenty before. The forest was open. Fruit trees, honey and grazing ground were abundant. Many pilgrims used to come from Nagpur to the Nagdwari temple to worship the Nag Baba. All used to stay in our village. We have heard that the pilgrims used to come on foot via Pachmarhi for the puja even before the British period. They still continue to come. There are no alternative routes. We do not have grain in our village for consumption throughout the year. Along with cultivation, we also prepare clarified butter. We collect honey and long grass from the forest. The grass is used for making brooms. We sell these in the market to earn a living. We also sell these things to the pilgrims who come for the Nagpanchami festival. We also bring firewood from the forest for cooking and keeping our homes warm during winter. We even sell it to the pilgrims who come for the Nagpanchami festival who use it for cooking purpose. One small bundle of firewood is sold for five rupees. The pilgrims stay in our house paying ten rupees per person per day. All the houses in our village have long patios. In each patio, about 30-40 persons can sleep. This work continues for only about two weeks. A bottle of honey fetches hundred rupees. We bring the honey from the hills. It is a tedious job but we have to do it to survive. We work as a team. Each team comprises of ten persons. It takes two days to bring the honey. Every person of a team get equal share.

A person gets a minimum of about four cans each. We sometimes get work in the forest to cut the lantana weeds or for making line to prevent forest fire. We get fifty rupees per day wages for the work. We even get to work as porter to carry the luggage of the people of the Nagpur Ashram and the pilgrims. There are opportunities to earn in the fair. The villagers set up stalls to sell tea, beetel nuts, bidi, chewing tobacco and puja materials. We also earn by carrying people and their luggages up the Nagdwari Hill who are physically challenged. We also earn by acting as guide on the Nishangarh Hill. The Mahadeve Mela Samity and forest deparment have constructed a dam for the pilgrims which is greatly beneficial for us. We get enough water for drinking and irrigation. With the establishment of the park, there is now restriction in bringing materials from the forest and agricultural land has also become less. The population in our village has also increased. There is also restriction in the construction of new houses and repair of the old ones. If a pilgrimage like Nagdwari has not been there, we would have been wiped out by now. What is there in a village? Nothing. There are no schools, no roads and no facilities for treatment. We do our own medication. We use herbal medication of the forest. Whenever we go to Pachmarhi, Bhopal or Nagpur and see the facilities available over there, we regret the difficulties we face in our village. To live in the city, you require education and intelligence. To get facilities, you require a job, which we don't have. If the government and forest deparment desires, they can establish good roads, hospitals and schools, revoke the restriction on collection of honey and grass from the forest and declare Nagdwari, Chitrashala, Nishangarh, Kalajhar as tourist spot and establish facilities over there, then our village will become the best. We can earn much more and take better care of the forest than what we do in the forest now and earn about ten thousand rupees every year. The fair that is being held from the time past can be done in a better way. Unpolluted air, water, hilly life style and pilgrimage will make our village a good tourist spot. It would be beneficial for all. What is happening at present is not right. To impose restriction of entry into the forest on those who stay in the forest is creating a gulf"

Case Study 8: Village – Rorighat
Informant: (1) Shri K. Lal (55 years, Male)
(2) Shri M. Singh (38 years, Male)

"Yahan kheti kam hai. Kheti ke upaj bhi kam hote hai. Jeene ke liye mehanat majduri karna padta hai. Woh bhi kabhi kabhi milta hai. Bachelog gaon ke school se aage pad nahi pate. Kahi aane jaane mein bahut duur tak paidal chalna padta hai. Suna tha, sarkar jangal ka bahar koi achcha jagah me le jayega jahan achcha kehti bari hoga, jeene ke liye aur bahut achche sadhan hoga. Ab tak is bare mein koi kam nahi hua hai. Lagta hai yeh sab baat kuch kam ke nahi hai".

"Agriculture is less here. Agricultural produce is also less. We have to work as labourers for our sustenance. That work also we don't get regularly. Our children cannot study beyond the school in the village. We have to walk a long way to go anywhere. We had heard that the government would rehabilitate us in some place outside the forest where there would be good agriculture and better facilities for living. Till date nothing has happened in this regard. It seems these are worthless talks".

Case Study 9: Village – Pattan
Informant: (1) Shri P. Singh (65 years, Male)
(2) Shri M. Ram (60 years, Male)

"Hamare gaon mein nadi ke kinare kheti ka jamin achcha hai. Paidawar bhi achcha hai. Sarso, chana, gehu ke saath saath bahut kisim ka sabji bhi ugaya jata hai. Sichai ke liye sarkar ke taraf se pani tanki banaya gaya hai. Nadi ke pani bhi pump karke jamin mein diya jata hai. Gaon ke kuch log fishing co-operative ke member hai. Woh log chhota chhota nau leke nadi mein machhli pakadte hai aur use co-operative mein bech dete hai. Achcha kamai hota hai. Co-operative woh machhli bazaar mein bechne ka bandobast karta hai. Hamar gaon bahut sundar hai. Isko chhod ke kahi nahi jayenge".

"We have good agricultural plots on riverside in our village. Agricultural produce is also very good. We cultivate mustard, gram, wheat and different types of vegetables. The government has erected water tanks for irrigation. We pump the river water also for harvesting. Some of our villagers are members of the fishing co-operative. They fish in the river using their small boats and then sell the catch in the co-operative. They earn well. The co-operative arranges to sell the fishes in the market. Our village is very beautiful. We would not leave it to go elsewhere".

(b) Nature of anthropogenic pressure in the core area

Density of population in the core area is a little below than 3 individuals per sq. km. As it appears, about 140 years before, ancestors of these communities were engaged for the destruction of forests. Since then till 1977, the same process continued. This has caused gradual erosion in their traditional knowledge regarding conservation of forest. When interviewed during survey, the local informants did not show much interest on this point. At present, they consider forest as the property and responsibility of the National Park Authority.

Cultivation, shifting cultivation (Dhaiya), rearing of cattle, hunting, fishing, collection of forest produces, etc., was their traditional occupation. With the formation of different protected areas in 1977, situation has changed. Shifting cultivation, hunting, fishing and collection of forest produces have been banned inside the Satpura National Park by law. The forest department has also taken away their agricultural lands. All these factors forced the villagers to face acute economic hardships. At present they practice cultivation, collection and selling of fuel wood from the forest, collection of non-timber forest products from forest, e.g. honey, mohul leaves, fruits, etc. A good number of the men go to nearby urban areas to work as daily wage labourers. A few of them are also in service both in government as well as in private organizations. Due to poor quality of soil, agricultural yields are low. So most of them depend totally on the forest for their survival. It was not possible to find out how much non-timber forest products are being annually collected from SNP. The villagers carefully avoided our questions and they are very scared of any government officials. However, it was reported that there are agents at Pachmarhi town and Matkuli who purchase their collections. Even these agents give them money on loan when necessity arises. However, from our observation and photographic documentation, it was possible to calculate about approximate quantity of fuel wood, which is being collected from the forest. It was reported that an individual goes to forest every alternate day and move around 6 to 7 km inside the forest to collect fuel wood. A head load of fuel wood varies from 20 to 25 kg. For their domestic consumption they require about 40 kg fuel wood per month. Rest of their collection is sold at Pachmarhi at the rate of Rs.20/- to Rs. 25/-. At Bariaam village, 58 individuals are engaged in this. Similarly at Pagara Dhana and Ghoranar villages, there are 56 to 52 persons respectively who have the same occupation. We checked the head loads carefully when they were being carried to Pachmarhi for selling and took photographs. It appeared that perhaps they just do not collect dried branches or dead wood. From the circumference of the collected fuel wood it appeared that these might be remnants of dead trees about 5 to 10 years old. At Binora village there are five families who distill liquor by fermenting mohua flowers. They distill it once in a week and sell their product at the weekly market at Bhurabhagat. During Shiv Ratri (mid February – mid March) they prepare mohua liquor for a month continuously to meet up the demands of the pilgrims who happen to pass through their village. As reported, on an average they distill mohua liquor for about 74 days in a year. Each indigenous distillation plant requires 10-15 kg of fuel wood at one go. Similarly at Kajri village, 70 persons collect firewood from the forest and store them. Besides domestic consumption, they sell the rest of their collection to the pilgrims during Nag Panchami (mid June – mid July) festival. They also go to the forest on every alternate day for this purpose.

Consumption of Fire/Fuel Wood/Mohua

1.	Domestic consumption (all villages)	40 kg x 259 families x 12 months = 1,24,320 kg per annum approx.
2.	Selling of firewood at urban centres	20 kg x 166 individuals (Bariaam, Pagara Dhana and Ghoranar villages together) x 180 days = 5,97,600 kg per annum approx.
3.	Selling of firewood at Kajri village during Nag Panchami	20 kg x 70 persons x 180 days = 2,52,000 kg per annum approx.

4.	Distillation of mohua liquor at Binora village	10 kg x 5 families x 74 days = 3,700 kg per annum approx.
	Total	9,77,620 kg per annum approx.

Besides they use wood for various other purposes. All these taken together, roughly it is estimated that about 1000 metric tons of firewood are being collected from an area of 524 sq. km, i.e., a little less than 2 metric tons are being collected per unit sq. km. We have no idea about rate of natural regeneration of vegetation inside the PBR core area where most of the pockets show relatively young trees and the vacant areas being infested with lantana shrubs.

In our opinion if this situation continues for another two decades, PBR core area would face irreparable damages to its existing fragile ecology. The Madhya Pradesh state government has decided that the core villages of PBR would be rehabilitated outside the biosphere reserve area but no practical steps have yet been taken. In the name of 'joint forest management', eco-development societies have been formed in all villages, but for many years they are lying defunct. Measures taken for the prevention of soil erosion are also inadequate. Due to acute shortage of cultivable land the villagers very often encroach forest land for agriculture. Except Jambudweep village, this particular phenomenon was observed in all PBR Core villages. In Pattan, villagers were found to till the Denwa river bank for their Ravi (winter) crop. Construction of the 'Tawa' dam did not appear to the survey team to be very rational, because its reservoir restricted free movement of the wild fauna inside PBR Core area. The Denwa river, which is an important contributor to the Tawa reservoir, every year carries a considerable amount of loose soil due to that agricultural activity on its bank. This would gradually elevate the level of reservoir bed and would cause flood in the area. If flood occurs, then fauna in the Madai sector of the PBR Core area would be immensely affected.

In the study area all agricultural fields were encircled by heavy fences in order to protect them from the wild herbivores. Usually thorny shrubs are used for this purpose. Due to local soil condition, forests of PBR Core area do not have much undergrowth. Even though whatever naturally grows there gets cut down by the villagers to erect fences. Denudation of undergrowth leads to loss of cover to the top soil and soil erosion.

Domestic cattle and poultry bird are highly valued in all the villages. They provide economic security to their owners in case of emergencies. Selling prices of of the same in market, during our survey, were:

(1) Hen - Rs. 100/- to Rs. 150/-
(2) Goat - Rs. 800/- to Rs.1200/-
(3) Cow - Rs. 5000/- to Rs. 10000/-.

During survey period it was observed that several big herds of domestic cattle were freely grazing inside the core forest nearer to each revenue/ forest village. Grass cover of the top soil as a result gets eroded, which in return causes soil erosion during each rainy season.

(c) Condition of tourism in Pachmarhi

The important tourist spots of Pachmarhi are: Priyadarshini Point, Handi Khoh, Apsara Vihar, Rajat Prapat, Rajendra Giri, Lanjee Giri, Jatashankar Temple, Chhota Mahadeo, Bara Mahadeo Cave, Chauragarh, Dhupgarh, Pandav Caves, Dorothy Deep, Dhuandhar, Harper Caves, Pachmarhi Church, Irene Pool, Twynam Pool, Duchess Fall, Sunder Kund, Ramy Kund, Bee Fall, Vanashri Vihar, Phansi Khad, Pratidhani, Pachmarhi Lake, Bison Lodge, Vihand Drishya, Denwa Darshan, Nishangarh and Nagdeo Temple.

As stated earlier, Pachmarhi was famous as pilgrimage centre, even in the pre-British period. In the past pilgrims from all over India used to come to Tamia. From there they used to visit Pachmarhi on foot via Nandya. The local Gonds and the Mawasis used to help these pilgrims by offering food and shelter. They also used to act as guide for the pilgrims. In return they used to earn money. Due to this occupation, the local tribals always had very cordial relation with the outsiders. At present, number of pilgrims have increased manifold. But their attitudes have also changed with the passing of time. Due to improved transportation system, better roads, etc., major share of the pilgrims directly go to Pachmarhi by bus or by hired cars from Tamiya or from Pipariya. The

Mawasis of Binora village, till date are helping these pilgrims who want to visit Chauragarh temple on foot during Shiv Ratri (mid February – mid March) festival. About six other tribal villages of the PBR buffer zone also unite together during the same festival. Fairs are organized at different places during this period. Modern agencies, both governmental and private, have entered into these affairs. Gradually the tribal people are loosing their control.

After formation of protected areas in Pachmarhi, adjoining areas of the township came under the control of State Forest Department and Satpura Tiger Reserve Authority. This arrangement of administration still continues. During this period to stop the damage of local ecology and forest, unplanned construction in and around the biosphere region and to stop the encroachment of outsiders into traditional tribal areas of the biosphere, some legal procedures were undertaken. Accordingly, Madhya Pradesh High Court issued a stay order. Due to that stay order any new construction or renovation of the old structures cannot be undertaken at present. Luckily our period of field investigation coincided with the biggest religious festival i.e. Shivratri, in the month of March. It was observed that continuous streams of pilgrims were pouring into Pachmarhi town by various modes of conveyance. The whole township was choking. Some pilgrims stayed in the dharamshalas, some in the open fields, some in the local hotels, and so on. It was very difficult to move inside the town. The pilgrims started to come about a week before the actual festival. The local authorities arranged a huge fair on the roadside. They auctioned the place to the local inhabitants for the fair, arrangement of electric connections, drinking water, food, entertainment, and so on. About 3 lakh pilgrims came during that festival. We were told that during Nagpanchami festival also, 2 to 2.5 lakh pilgrims come to this region. Due to lack of infrastructure these pilgrims usually suffer a lot, particularly those who find no place either in the dharamshalas or in hotels. These conditions create health hazard to the inhabitants living inside the core area villages. Pachmarhi town is encircled on three sides by the PBR core area. The Mawasi people told us that after these big festivals, they offer special prayer to Mailee Mata in the jungle so that no epidemic breaks in there because the pilgrims have hardly any proper sanitation facility. They usually stay near the hill streams. The human waste contaminates the hill streams, which is the source of drinking water for the local tribals. Besides these two festivals, Pachmarhi also receives tourists almost throughout the year. In our modest estimate 8 to 9 lakh of people visit Pachmarhi every year. Due to High Court's injunction and the restrictions imposed by the forest department, development of infrastructure with modern facilities cannot be undertaken though Pachmarhi has immense potential for the growth of tourism industry. The problem is that some of these temples and few tourist spots are located right inside the core area so during festivals lakh of people pass through the core area, which the State Forest Authority or the Satpura Tiger Reserve Authority cannot stop. This is a checkmate type of situation.

Proposals for Future Actions in the PBR

Duration of our survey was relatively short in the PBR core area. Though physical condition of the study area was very much challenging yet our team extensively travelled in all the corners of the core area as far as was possible and came up with an observation that, perhaps proper coordination with respect to planning, implementation of development programmes, etc., appears to be lacking among the education department, settlement department, forest department and the tourism department. As a result the tribal people living inside PBR core area are is stress and strain.

Under these circumstances, in our opinion, the following measures may be undertaken immediately:

1. Rehabilitation for the core area villagers may be done immediately.
2. Forest department may employ them in more numbers for the conservation of Satpura National Park, so that each adult individual may be able to get minimum 100 days working opportunity in a year.
3. Rehabilitation sites may be carefully selected where they can get good communication system, safe drinking water, higher secondary schools and primary health centers/ hospitals, along with employment opportunities.

4. Use of LPG may be made mandatory in the adjoining urban centers of the PBR core area. If there will be no demand for firewood, young plants of PBR would be saved.
5. Pachmarhi has immense potential for tourism. Proper planning may be undertaken for the construction of necessary infrastructures with modern facilities. Proper roads may be constructed so that tourists can visit all the tourists spots, mentioned earlier in this summary report. In each case entry fee may be levied. Amount of money thus collected may be used for the maintenance of infrastructures, conservation of ecology in the Pachmarhi Biosphere Reserve and development of entire Pachmarhi area.
6. Proper plans may be made to remove all the slum type settlements from Pachmarhi town.
7. Souvenir shops may be opened near all tourist spots for selling of picture post cards, tribal handicrafts and forest products like honey, *chironjee*, etc.

2

Management of Natural Resource and Practice of Indigenous Knowledge in the Core Area of the Pachmarhi Biosphere Reserve

Samit Ghosal

Pachmarhi Biosphere Reserve (PBR)

Pachmarhi area was designated as Biosphere Reserve by Govt. of India vide its Notification No. J-220116/17/94-BR, dated 03rd March, 1999 on the basis of project Document prepared by Environmental Planning and Coordination Organization (EPCO) an autonomous body of the Govt. of Madhya Pradesh. The state government has also appointed Environmental Planning and Coordination Organization (EPCO) as the Nodal Agency for implementation of Management Action Plan of Pachmarhi Biosphere Reserve.

Pachmarhi Biosphere Reserve beside being a vast natural resource has diverse flora and fauna and a unique topography. It is also the home of a sizeable number of tribal population.

The Pachmarhi Biosphere Reserve lies in between latitude 22°11′ to 22°56′ N and 77°47′ to 78°52′ E longitude and covers three civil districts of Hoshangabad, Betul and Chhindwara of the state of Madhya Pradesh. The total area of the Pachmarhi Biosphere Reserve is 4987.38 sq. km. The PBR envelops three wildlife conservation units, viz. Bori Sanctuary (485.72 sq. km.), Satpura National Park (524.37 sq. km.) and Pachmarhi Sanctuary (417.78 sq. km.). All the three wildlife conservation units, Satpura National Park, Bori Sanctuary and Pachmarhi Sanctuary, comprising an area of 1427.00 sq. km., have been declared as Project Tiger Reserve in August, 2000. In Pachmarhi Biosphere Reserve, the Satpura National Park (524.37 sq. km.) is considered as the Core Zone and the remaining area of 4525.93 sq. km. including Bori Sanctuary (485.72 sq. km.) and Pachmarhi Sanctuary (417.78 sq. km.), serve as the buffer zone of the PBR. As such, so far there is no demarcated transition area in the PBR. (Source: EPCO PBR Documents, Vol. 1 (No.1), 2001)

The land use pattern of the area based on landsat imagery of April, 1990 and November-January 1991-1992, comprises of land forest (65.2%), agricultural land (27.7%), water bodies (4.2%), wasteland (2.5%) and buildup land (0.5%). Of the total forest cover, the closed forest constitutes 85.3%, open forest 8.2%, degraded forest 4.2% and forest blank 2.3%. Tawa reservoir is the major constituent of the water bodies. Tawa and Denwa are the two main rivers in the area. Preponderance of Sal tree (*Shorea Robusta*) in the predominantly teak bearing area is a unique feature of Pachmarhi Biosphere Reserve.

Bori Sanctuary is the oldest reserve forest in India. The Bori Reserve forest was declared as a Wildlife Sanctuary in 1977 under the Wildlife Protection Act of 1972 by the Govt. of Madhya Pradesh vide Notification No. 15/22/76/X(8) of dated 01/06/1977. Pachmarhi Sanctuary was created out of Bori Sanctuary in the year 1977 and then Satpura National Park was created out of Bori and Pachmarhi Sanctuaries in the year 1981, vide Govt. of Madhya Pradesh Notification No. 15/12/80-X(2), dated 13/10/1981.

The Pachmarhi Biosphere Reserve area comprises of 622 villages (Source: EPCO). The area is largely an agricultural rural setting with preponderance of forest indicative of its hinterland situation. Agriculture is the mainstay of the population. Out of the total population of 2,17,820 (1981 Census), the scheduled castes account for 14.3% and the scheduled tribes account for 41.5%. The average family holding is only 3-4 hac. The literacy rate is low (21%). The cattle and goats are reared in large numbers in the area due to the presence of vast forest land available for pasture and grazing. Agriculture is the main source of income. Working in the forest area is the supplementary source of income for the village lying in the vicinity of the forests. The village houses are generally constructed with mud and wooden poles. Common Hindu and Muslim festivals are observed.

Indigenous Knowledge

The indigenous knowledge system present among the tribal people of the area needs to be urgently addressed because the bio-diversity present in the Pachmarhi Biosphere Reserve is a big repository of human's traditional knowledge base. The indigenous knowledge of the people can make valuable contribution to the modern and contemporary knowledge. Biosphere Reserves are seen to be the areas where traditional knowledge of the tribal people can be preserved and enhanced.

Indigenous knowledge is a set of perceptions, experiences and behaviours that guide the local people in the use of land and natural resources. Tribal (or indigenous) people inhabiting the area possess an immense knowledge of their local ecosystem based on years of living close to nature. The technique of using and managing their natural resources, such as, land and forest for foods, materials for house construction and making of agricultural tools, medicines and other products. These experiences are cumulative and dynamic in nature and are passed on from one generation to another. It evolves in response to changing ecological, economic and socio-political circumstances. Along with conservation and management of biological diversity, the recognition and production of this knowledge and sharing of equitable benefits derived from the use of this knowledge with its holders is the need of the hour. Indigenous knowledge is a fundamental component of indigenous culture – its products and expressions or actions like agricultural operations, subsistence pattern, land management, cultural performances, beliefs, ceremonies and ritual performances, are all integral part of it.

Following are the Salient Parameters of the Indigenous knowledge –

1. Unique to a given community
2. Based on cumulative experience

3. Often tested over centuries of use
4. Adapted to local cultures and environment
5. Dynamic and changing with conditions
6. Transmitted through indigenous communication channels

Objective of the Study

Pachmarhi Biosphere Reserve, with its three conservation units, viz. Satpura National Park, Pachmarhi Sanctuary and Bori Sanctuary, is considered as a protected area. There are all together 73 villages within these conservation units, of which 26 are forest villages and 47 are revenue villages. Satpura National Park has ones forest village and six revenue villages, Bori Sancutary has all 17 forest villages and Pachmarhi Sanctuary has eight forest villages and 41 revenue villages. These villages of the protected area within the Pachmarhi Biosphere Reserve (PBR) have also become part of Satpura Tiger Reserve (STR) under project Tiger, the main funding agency for the area. The people living in these villages are cocooned within layers of statutory protection units and are subjected to a series of prohibitions and protective measures imposed by forest department, mandate of Biosphere Reserve and Tiger Reserve Authority. Two conflicting approaches are evident in the management of the villages in the Pachmarhi Biosphere Reserve area. The people oriented doctrine of Biosphere Reserve recognises some role for the local people at least outside the core zone. This is in total contradiction with the dialectic principles of total protection of wildlife in parks and sanctuaries without any human interference whatsoever in the Tiger Reserve area.

In the wake of this traumatic situation it would be interesting to note how the people living in the core zone of the Pachmarhi Biosphere Reserve are adjusting to this imposed condition. Their adaptation to natural resources with special reference to the exploitation and management of land and forest. Subsistence economy of the people and application of their local and indigenous knowledge in relation to their livelihood pattern would be studied. Traditional techniques used to overcome or cope with the crisis pertaining to health. The inhibiting or facilitating factors in the conservation are use of traditional knowledge and the techniques.

The areas of enquiry pertaining to the Management of Natural Resources and Indigenous knowledge would be –

1. Land use pattern
2. Agriculture
3. Crop rotation
4. Soil and water conservation
5. Production of crop
6. Land and forest Management
7. Use of medicinal plants
8. Marketing of forest produces
9. Relationship of human interface with the Biosphere

The main thrust of enquiry would be centered around the (i) Role and activities of the tribal people living in the core area villages of the Pachmarhi Biosphere Reserve in relation to their natural environment (ii) Cultural values and knowledge system of the people vis-à-vis their environment and (iii) How far the traditional knowledge in managing their eco-system could be incorporated into the modern system of production to obtain a synergy.

Area of Study

In the Satpura National Park, the core zone of the Pachmarhi Biosphere Reserve, there are seven village settlements. Six are complete villages and one is a hamlet of a village, which comes inside the National Park. The other hamlet of the village remains outside the park and core zone, hence is not considered for the study.

The villages which have been covered under the study are as follows –

Table 2.1

Sl. No.	*Name of Village*	*Location*	*Type*
1.	Ghodanar	Satpura National Park	Forest Village
2.	Dhana (Part of Pagara)	Satpura National Park	Revenue Village
3.	Badkachar	Satpura National Park	Revenue Village
4.	Kajri	Satpura National Park	Revenue Village
5.	Rorighat	Satpura National Park	Revenue Village
6.	Binora	Satpura National Park	Revenue Village
7.	Chitor Pattan	Satpura National Park	Revenue Village

These villages situated within the Satpura National Park are at a great distance from each other. Locations of each village are unique and so are their soil type, topography and ecological surroundings.

Methodology

Taking village as a unit of study, all the seven village settlements in the core zone of the Pachmarhi Biosphere Reserve/ Satpura National Park have been studied by using standard anthropological tools and techniques. Data were collected through case studies, personal interviews as well as open-end questionnaires, schedules and personal observation. The fieldwork for the study was conducted during the month of February and March, 2005.

Villages under Study

Data were collected from the seven villages in the core area of the Pachmarhi Biosphere Reserve. The entire area also comes under Satpura National Park. Village boundaries are called *bagar*, pillars called *munara* that are erected by the forest department on the boundary lines separates it from the forest land. All the villages in the core area of Pachmarhi Biosphere Reserve, excepting Pattan, are almost exclusively populated by the Mawasi community, Pattan village is inhabited by the Gond, popularly called Gond-Thakur.

The houses are indigenously made by using local materials. Fences in the homesteads are called *bagar*. They use strips of *tatta* bamboo for the fence, woven in lattice design. House walls are made of indigenously made earthen bricks. The bricks are made by mixing clay and sand, put into wooden crust and sun dried. They use bamboo, *sonta* grass and *sukra* grass for thatching of roof. White soil or *churi mitti* is used for plastering of walls. *Lal mitti* (red ochre) and *pila mitti* (yellow soil) are used for flooring. Measurement of land is called *jarip*. A 50 feet long chain is also called *jarip*. Possessing 4 *jarip* of land, means, a piece of land that is 200 feet (4 X 50 ft) in length and 200 feet in breath, i.e. more or less one acre of land.

A brief note on the villages are given below:-

1. Ghodanar – Ghodanar is a forest village situated at a distance of 22 km from the Pachmarhi town. It is located within Satpura National Park and administered by the forest department. The settlement pattern of the village is linear type with houses situated on either side of the main village road. All the land of the village belong to the Mawasi, a scheduled tribe community. Ghodanar has 27 families.

2. Dhana – Dhana is a revenue village, rather a hamlet of the Pagara village, which together with Pagara is called Dhana-Pagara. The Pagara section of the village lies out of the National Park and is in the Pachmarhi Sanctuary, which has not been considered for study – Dhana is situated at a distance of 16 kilometers from Pachmarhi town.

3. Badkachar – It is a revenue village situated well inside the forest of Satpura National Park. It has three hamlets and the main settlement is called Badkachar. This is a big village, diverse in nature but circular in shape. The people can not fully use the agricultural land in their possession due to scarcity of water and rough terrain of the land which is full of stones and boulders. Good lands in the village are in possession of only four to five families. Most of the families are land less in the village. It is situated at a distance of 16 kilometers from Pachmarhi town.

4. Kajri – Kajri, a revenue village, is situated at a distance of 25 kilometers from Pachmarhi town. The village is located in a deep valley with steep hills surrounded on all sides. It is famous for the Nagpanchami fair held in the month of Shravan. Hundreds of pilgrims come to *nagduari* cave near Kajri for worship. The inflow of religious tourism has boosted the economy of the village. Local people earn a lot by working as porters for the tourists and they also provide for the food and lodge (accommodation) for the pilgrims. The people use their agricultural lands as shelters for the tourists by erecting temporary sheds. The fields near the fair ground in the village do not produce any *kharif* crops during monsoon as it is engaged for tourism purpose. The porters get Rs.200/- for a luggage weighing 30 kilogram. For 20 kilogram they get Rs.100/-. Carrying of pilgrims' luggage is called *bojha dhona*.

5. Rorighat – Rorighat is situated in Satpura National Park at a distance of 18 kilometers from the Pachmarhi town. In Rorighat houses are surrounded by agricultural fields called *bari*. In Rorighat a class of land called *rajos* jungle is available for slash and burn cultivation. These are vested revenue land outside the forest limit which are occupied by the villagers for cultivation.

6. Binora – Binora is located just in the border of Satpura National Park adjacent to the Chindwara district. It is situated at a distance of 106 kilometers from the Pachmarhi town. The Denua River in the village separates Hoshangabad district from Chindwara. The village serves as the gateway for the pilgrimage to Mahadeo cave and during *Sivaratri* i.e. 08th March, hundreds of pilgrims visit the village on their way to the Mahadeo fair. Selling of mahua liquor to the tourists is most common in Binora. The Denua river bed is wisely used for the distillation of mahua.

7. Pattan – Pattan or Chitor Pattan village unlike the rest of the villages in core area is inhabited by the Gond. It is located inside the Satpura National Park. In its nearby forests lots of wildlife is found. The Denua river is very wide here and flows by the side of the village. During winter when the water level recedes, a sizeable amount of land with very fertile soil are available for cultivation on both sides of the bank. These lands are called *dub jamin*. Rabi crops are cultivated in these lands. It is situated at a distance of 126 kilometers from the Pachmarhi town.

Demography

Demographic detail of these villages were collected from the village itself, as well as from the secondary sources. However,

Table 2.2
Population Structure (Census: 2001)

Sl. No.	Village	Persons	Male	Female	Sex Ratio
1.	Badkachar	155	75	80	1067
2.	Binora	200	92	108	1174
3.	Ghodanar	111	62	49	790
4.	Kajri	162	86	76	884
5.	Rorighat	234	120	114	950
6.	*Pagara / Dhana	--	--	--	--
7.	Chitor Pattan	206	115	91	791

**** Separate data for Dhana is not available as it is clubbed together with Pagara.***

Table 2.3
Family Size (Census: 2001)

Sl. No.	*Village*	*Nos. of Family*	*Persons*	*Family Size*
1.	Badkachar	26	155	6.0
2.	Binora	32	200	6.3
3.	Ghodanar	19	111	5.8
4.	Kajri	28	162	5.8
5.	Rorighat	42	234	5.6
6.	*Pagara/ Dhana	--	--	--
7.	Chitor Pattan	44	206	4.7
	TOTAL	**191**	**1068**	

**** Separate data for Dhana is not available as it is clubbed together with Pagara.***

Table 2.4
Population Structure (Field Data: 2005)

Sl. No.	*Village*	*Nos. of Family*	*Population*	*Family Size*
1.	Dhana	23	127	5.52
2.	Ghodanar	17	114	6.70
3.	Badkachar	31	197	6.35
4.	Rorighat	45	256	5.68
5.	Kajri	29	187	6.44
6.	Binora	40	239	5.97
7.	Chitor Pattan	47	237	5.04
	TOTAL	**232**	**1357**	--

some discrepancies are noticed between the field data and available secondary data. In the following pages the available demographic data are presented in tabular forms.

Management of Natural Resource

Exploitation of natural resources is a precondition to human survival and subsistence. Peoples living in different villages in the

Satpura National Park i.e. the core zone of the Pachmarhi Biosphere Reserve manage their natural resources in a sustainable way by adapting to the local conditions. Geomorphological condition, topography, soil types etc, are different in each village and the people living therein adapt various techniques for the utilization of natural resources. They also improvise it according to the situation and demand. The Dynamic nature in its adaptability is very much evident as far as the management of natural resources are concerned.

Land and forest are the two main natural resources available to them for exploitation and they manage it in the traditional or indigenous way which is imbedded in their culture.

Management of Land Resource

There are seven villages located within the Satpura National Park/Core Zone of the Pachmarhi Biosphere Reserve. One village i.e. Ghodanar is a forest village, and rest are revenue villages. All these villages are inside the protected area of Satpura Tiger Reserve (STR). Ideally, all these villages are to be relocated out side the Satpura Tiger Reserve (STR) and core zone of Pachmarhi Biosphere Reserve. The process of relocation has already been initiated by the forest department by annexing the cultivable land into the forest. Now the villages are left with very little amount of land for agricultural purpose. In this meager amount of land it has become very difficult for the people to support the population as their main source of livelihood being land. They adopted various devices for the management and maintenance of subsistence economy.

In the Table 2.5, it is shown that how much land the villages had to forego to the forest department as of now. To compensate the loss they started illegal encroachment of forest land in some places, at places they cultivate the river bed and in some other places they have taken possession of vested revenue land and use it for slash and burn cultivation adopting an indigenous method. Ghodanar being a forest village the right of land belongs to forest only. In following table, state sponsored land alienation is shown which is effective in the core area villages of Pachmarhi Biosphere Reserve after the declaration of Tiger Reserve.

Table 2.5
Land Transfer to Forest Department

Sl. No.	*Village*	*Original Area of Revenue Land (in hac.)*	*Area Transferred to Forest (in hac.)*	*Area for Cultivation (in hac.)*
1.	Rorighat (R.V.)	829.60	614.71	214.89
2.	Kajri (R.V.)	1005.75	991.12	14.62
3.	Badkachar (R.V.)	1854.42	1813.41	41.00
4.	Binora (R.V.)	764.20	702.50	61.69
5.	* Dhana (R.V.)	--	--	--
6.	*Ghodanar (F.V.)	--	--	--
7	* Pattan	-	-	-

Source: Records – Forest Department
* Figures not available.

From the above figures it is evident how the agricultural lands, once belonged to the village, are alienated to the state for some greater goal. However, this nonetheless has opened up new avenues of livelihood subsistence for the villagers.

With this meager land it has become very difficult for them to support their livelihood, as their main economy was land based subsistence agriculture. In response to this situation they have adapted various subsistence strategies to compensate the loss of land. However these strategies are different in different villages, which are as follows:

1. Illegal encroachment of forest land for cultivation
2. Cultivation of rabi crop in the river bed when there is no water
3. Utilisation of vested revenue land with the help of a unique slash and burn type of cultivation.
4. Intensify collection of forest produce for commercial use such as honey,mohua etc.
5. Providing services to pilgrims to promote religious tourism.

Table 2.6
Land Owning and Land Less Families

Village	*Land owning Familes*	*Percent*	*Land less Familes*	*Percent*	*Total No. of Families*
Dhana	14	60.86	9	39.14	23
Ghodanar	17	100.00	0	00.00	17
Badkachar	13	41.94	18	58.06	31
Rorighat	27	60.00	18	40.00	45
Kajri	24	82.75	5	17.25	29
Binora	15	37.50	25	62.50	40
Chitor Paltan	38	80.86	9	19.14	47
Total	148		84		232

Land under Cultivation

Deliberation under the above topic is based on empirical data collected during field investigation. Here we will discuss the amount of land available for cultivation in each village and number of family/people depend on it for their livelihood from subsistence agriculture. Use of land for cultivation is seasonal in nature and varies from season to season. Some land in Kajri village is used as a shelter for the pilgrims coming for the *darshan* of *nagduari* and so is not in use for *kharif* cultivation this increases its productive power for *ravi* cultivation.

In the Table 2.7 we will show the per capita land available for utilisation for agricultural purpose in the villages under study.

Table 2.7
Per Capita Land Holding

Sl. No.	*Village*	*Population*	*Land (in acre)*	*Per Capita Holding (in acre)*
1.	Ghodanar	114	27.0	0.23
2.	Dhana	127	24.5	0.19
3.	Badkachar	197	62.5	0.31
4.	Kajri	187	30.5	0.16
5.	Rorighat	256	26.5	0.10
6.	Binora	239	57.5	0.24
7.	Pattan	237	131.5	0.55

Table 2.8
Family Wise Land Holding

Sl. No.	*Village*	*No. of Families*	*Land (in acre)*	*Average Holding (in acre)*
1.	Ghodanar	17	27.0	1.58
2.	Dhana	23	24.5	1.06
3.	Badkachar	31	62.5	2.01
4.	Kajri	29	30.5	1.05
5.	Rorighat	45	26.2	0.58
6.	Binora	40	57.5	1.43
7.	Pattan	47	131.5	2.79

From the above two tables it is evident that as far as the holding of land is concerned the people of Pattan are in best position and people of Rorighat are in worst position. Now we will see how far this affects the level of crop production in these two villages as well as in other villages also.

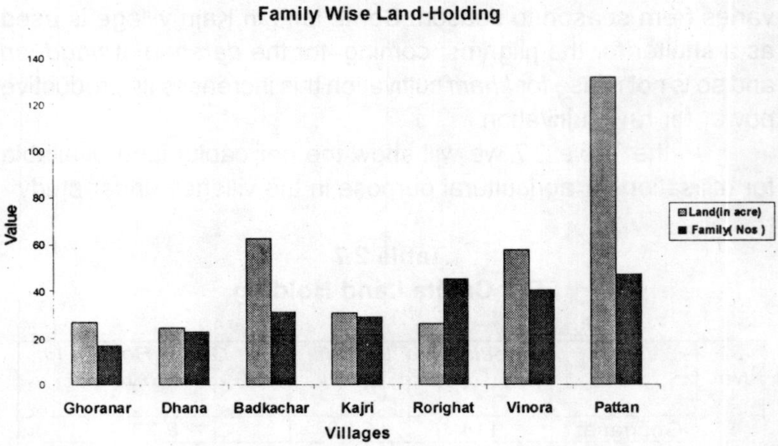

To draw a line of comparison between land holding and crop production in Table 2.9 we have shown the village wise per capita production of crops in the seven villages situated inside the core area of the Biosphere Reserve.

Production of Crops

Table 2.9
PER CAPITA PRODUCTION OF CROPS

Sl. No.	*Village*	*Population*	*Crops Produced (in Quintal)*	*Crop Production (per capita)*
1.	Ghodanar	114	73.99	0.64
2.	Dhana	127	54.47	0.42
3.	Badkachar	197	32.15	0.16
4.	Kajri	187	192.1	1.02
5.	Rorighat	256	168.7	0.65
6.	Binora	239	77.12	0.32
7.	Pattan	237	358.77	1.51

From the Table 2.9 it is observed that the lowest per capita production of crops is in Badkachar, whereas though in Rorighat both the per capita land holding and average per family land holding are lowest, but as far as the crop production is concerned it is not the lowest. Binora village though holds more per capita land than Rorighat but Rorighat is far beyond in production of crops. This is to due the presence of very infertile and stony land in Binora. The people of Rorighat by effective and efficient management of agricultural operation, crop selection and technique are able to grow more crops in comparative small patch of land.

Table 2.10

Per Capita Crop Production and Per Capita Land Holding

Sl. No	*Village*	*Per capita land holding (in acre)*	*Per capita crop production (in quintal)*
1	Ghodanar	0.23	0.64
2	Dhana	0.19	0.42
3	Badkachar	0.31	0.16
4	Kajri	0.16	1.02
5	Rorighat	0.10	0.65
6	Binora	0.24	0.32
7	Pattan	0.55	1.51

Here we can see that the lowest per capita production of crops is in Badkachar though in Rorighat both per capita land holding and average land holding are lowest but as far as the production of crops is concerned it is not the lowest. Binora though holds more per capita land than Rorighat, is far behind in production. It is also seen that in Pattan per capita production of crops goes along with per capita holding of land. Whereas in Rorighat though the per capita land holding is least its per capita production of crops is more than many villages. In Kajri though the per capita land holding is only 0.16 acre, its per capita production of crops is 1.02 which is only next to Pattan.

There are various factors which are responsible for this unique unequal scenario of crop production. Due to its infertile and stony land people of Binora have concentrated on mohua collection and they sell the liquor during Mahadeo fair in Sivaratri. The people of Rorighat by effective and efficient management of agricultural operation , crop selection and rotation are able to grow more crops in comparatively small patch of land. The excess cash money received by the people of Rorighat during Nagpanchami are used in buying chemical fertilizer for their agricultural fields which in return gives them a good production. Some fields in Kajri are left fallow during kharif season as they are to attend to the work of the pilgrims visiting for Nagduari mela.

Cropping Pattern

Yield or productivity of the land is an important phenomenon to understand and evaluate a subsistence agricultural economic system. One of the ways of measuring the output – how much is produced in a standard area such as per acre – over a specified period of time, usually a year, to correspond the outputs of temperate regions where the annual variation in weather generally results in a variation in yield in single crop season. The yield of a particular crop can be measured in weight when one community grows several crops, the productivity of one may be much more than that of another – more weight, more bulk, more calories and more people fed. In order to understand the management of land as prime natural resource, the interrelations of environment, agricultural practices and land use are to be carefully examined. We have compared the rate of production in the core area villages of Pachmarhi Biosphere Reserve where the same crops or at least the same type of crops are grown. The difference in the production of same crop by the same community in two or more places accounts for the variations in the soil type, technology involved, manure used and necessity of that crop by the people along with geo-morphological conditions are the vital factors for the production of crops. A vital aspect of food production in the villages under study is that there are large differences in the quantities of food crops raised by different villages under study. As a result, different villages tend to produce more of certain crops.

Table 2.11
Production of Crops in Village
(Pachmarhi Biosphere Reserve: Core Area)

(In Kgs)

Crops	*Village*						
	Ghodanar	*Dhana*	*Badkachar*	*Kajri*	*Rorighat*	*Binora*	*Pattan*
Makka	2627	1870	1490	2087	4070	3626	1262
Wheat	2760	2240	1050	15880	4700	1600	12500
Paddy	1012	455	251	850	00	400	7950
Juari	505	197	50	235	100	347	00
Kutki	24	00	00	00	7880	400	500
Kodon	00	00	00	00	00	1250	00
Samara	00	00	00	00	00	00	4405
Tuar	00	00	215	00	00	55	20
Urad	135	00	23	25	00	05	00
Chana	00	00	00	85	110	00	9160
Matar	00	78	00	00	00	00	00
Barbati	122	102	91	48	10	29	10
Soyabean	210	300	00	00	00	00	00
TOTAL	**7399**	**5447**	**3215**	**19210**	**16870**	**7712**	**35817**

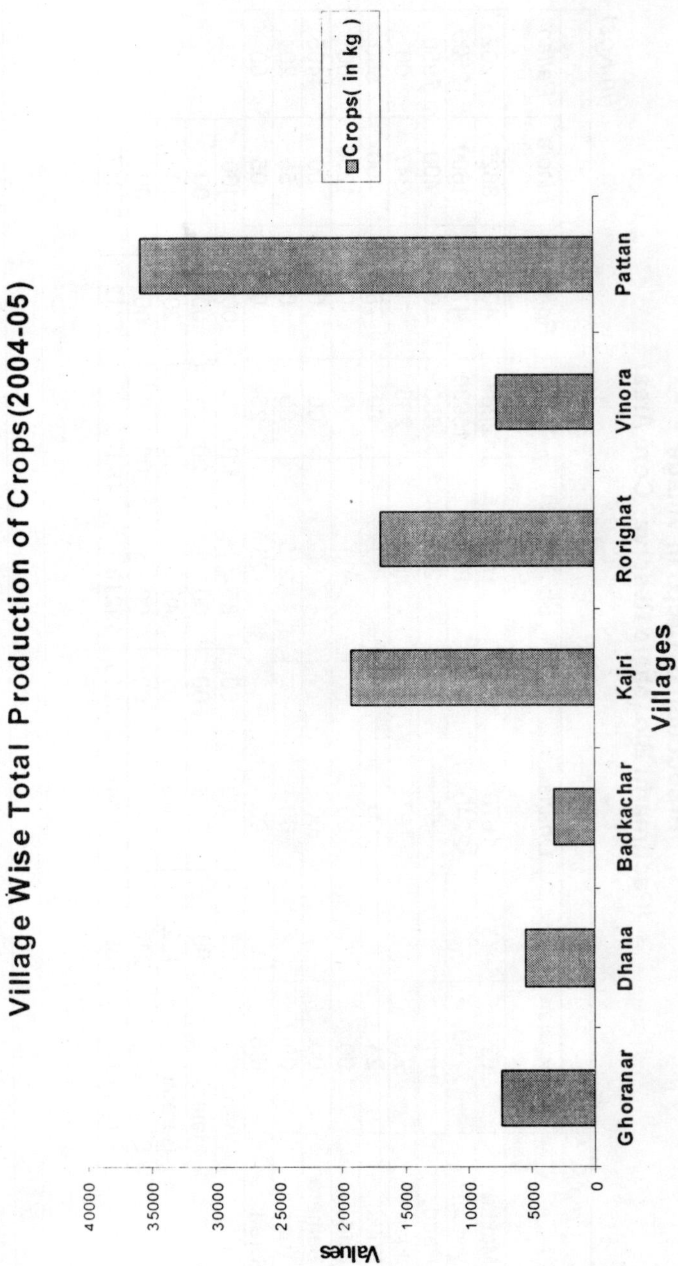
Village Wise Total Production of Crops(2004-05)
Crops(in kg)
Values
40000
35000
30000
25000
20000
15000
10000
5000
0
Ghoranar
Dhana
Badkachar
Kajri
Rorighat
Vinora
Pattan
Villages

Figure : 1

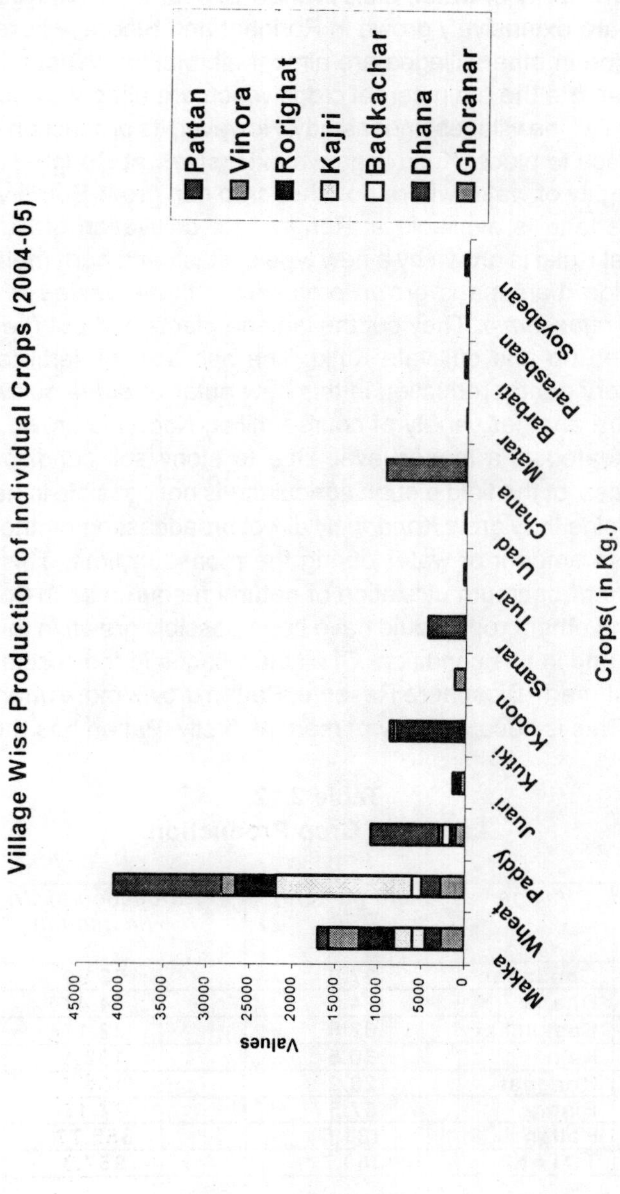
Village Wise Production of Individual Crops (2004-05)
Pattan
Vinora
Rorighat
Kajri
Badkachar
Dhana
Ghoranar
Values
45000
40000
35000
30000
25000
20000
15000
10000
5000
0
Makka
Wheat
Paddy
Juari
Kutki
Kodon
Samar
Tuar
Urad
Chana
Matar
Barbati
Parasbean
Soyabean
Crops(in Kg.)

Figure : 2

From the Figures 1 and 2 it is observed that some villages have special preference or choice for a particular type of crop due to the inherent ecological conditions like soil types, elevation of field, availability of water, etc. A variety of coarse Millet, Kutki and Kodon are extensively grown in Rorighat and Binora, whereas its cultivation in other villages are almost nil. Makka, Wheat, Paddy and Juari are the main cereal crops which are almost exclusively grown in all the villages under study. However, its production varies form place to place. Kutki is grown extensively at Rorighat due to the scarcity of water where no other crop can grow. Some vested revenue land is available at Rorighat for cultivation of Kutki. At Rorighat Kutki is grown by a new type of slash and burn method of cultivation. Lantana is grown profusely in these vested revenue land or *rajas jamin*. They cut the lantana plants and put fire when it is dried up and cultivate Kutki. The ash acts as fertilizer and gives very good production in this new abtar of *dehia* cultivation. At Binora another variety of course millet, *Kodon* is grown in the land situated at a higher level. Due to stony soil condition and steepness of the field plough agriculture is not possible in here. In these fields they grow *Kondon* by direct broadcasting method with very less amount of water during the monsoon time. This is an example of optimum utilization of natural resource as in both the cases no other crops would have been possibly grown in this type of land and in this condition. Of all the villages in the core area of the Pachmarhi Biosphere Reserve, Pattan grows more quantity of crops. This is because of two reasons firstly, Pattan has the best

Table 2.12
Land and Crop Production

Sl. No.	*Village*	*Land (in acre)*	*Production of Crop (in quintal)*
1.	Ghodanar	27.0	73.99
2.	Dhana	24.5	54.47
3.	Badkachar	62.5	32.15
4.	Kajri	30.5	192.1
5.	Rorighat	26.2	168.7
6.	Binora	57.5	77.12
7.	Pattan	131.5	358.77
	TOTAL	**359.7**	**957.3**

agricultural fields, plain and good soil and secondly, it has some river side fields available for cultivation during summer and winter which are very fertile due to the silt deposition in monsoon. This type of field at Pattan is called *dubjamin*.

Production of crops is always related to the amount of land available for cultivation in a particular village at a given point of time. By comparing the land available under cultivation with the quantity of crop produced, we can get an idea on the productivity of that village.

From the Table 2.12 it is observed that land holding wise production is maximum in village Pattan. Land holding wise production is lowest in Badkachar.

Increased rate of production of crops depend on various parameters like, soil type, technique of cultivation, irrigation etc., one major parameter being the use of chemical fertilizer. Pachmarhi Biosphere Reserve villages use Urea and DAP in their field only for the cultivation of Wheat, for the rest of the crops they mainly rely on organic manure like Cow-dung.

Table 2.13
Fertilizer Used

Sl. No.	*Village*	*Land (in acres)*	*Fertilizer Used (in Quintals)*
1.	Ghodanar	27.0	1.55
2.	Dhana	24.5	0.5
3.	Badkachar	62.5	2.75
4.	Kajri	30.5	21.9
5.	Rorighat	26.2	7.27
6.	Binora	57.5	0.24
7.	Pattan	131.5	12.54
	TOTAL	**359.7**	**46.75**

Table 2.13 indicates the land holding wise use of chemical fertilizer in the core area villages of Pachmarhi Biosphere Reserve. It is seen that maximum amount of fertilizer (21.9 quintals) is used by Kajri in 30.5 acre of land which accounts for the villages high quantity of crop production (192.1 quintal) in the land. The people of Kajri use their income from tourism in the purchase of fertilizer which in turn gives them good production of crops.

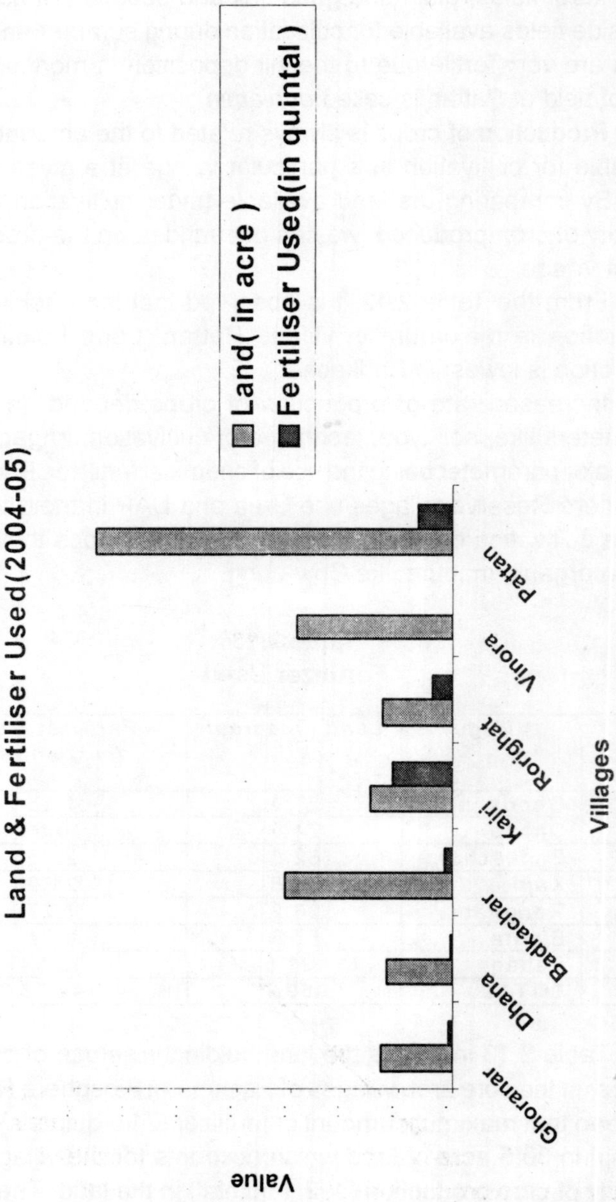
Land & Fertiliser Used(2004-05)
Land(in acre)
Fertiliser Used(in quintal)
Value
Pattan
Vinora
Ronghat
Kajri
Badkachar
Dhana
Ghoranar
Villages

Figure : 3

Agricultural Operation and Production

Under this topic, we will discuss the method of agricultural operation, selection of crops for cultivation, crop rotation and all those parameters used by the people of core area villages to obtain optimum result in output for effective and efficient resource management.

Type of Land and Soil

Basically, there are two types of land *Bare* and *Khetty*. *Bare* land is situated near homesteads and *Khetty* is located at a distance. *Bare* means Mango, i.e., where Mango grows. There are two to three types of soils. *Shera* is sandy loam soil. Makka (maize) is the best suitable crop for *Shera* though Wheat can also grow as a winter crop. *Motar jamin kali mitti* is the best soil type with more water retention capacity where even Paddy can grow. Paddy land is called *juban*. It is low lying muddy soil . *Motali motar* is the black soil with very good production. *Shera* is the mixture of red (loam) and White (Sandy) soil. But the ratio can be different, when the proportion of white soil is more, crop like Maize, Wheat, *Chana* grow better, but when it is more of red soil, crops like *Urad*, *Barbati*, Beans grow more. *Konjni* is the kitchen garden. In *motar* soil all types of crops can be grown with irrigation. In *Shera* soil cow dung is required for Paddy cultivation but in *motar* soil Paddy can be produced even without cow dung manure. In *motar* even DAP and urea are required very less for Wheat cultivation. In Pattan we have another unique soil type in the form of *dube jamin*. It is a very fertile stretch of land on the bank of river Denua. This land remains under water during monsoon. It is cultivated from the month of Maha to Chait/Baishak. *Chana* and Wheat are the main crops cultivated here. *Dub jamin* consists of sandy (*retali*) soil. Since the land remains under water, silt used to deposit in it and produce bumper crop without any fertilizer or manure.

Agricultural Calendar Eight and Months

Agricultural season in this area starts from the month of *Chait* and the last month of the local calendar year is *Baishak*.

Table 2.14
Local Calendar

Sl. No.	*Local Months*	*English Equivalents*
1.	Chait	Middle of March to Middle of April
2.	Baishak	Middle of April to Middle of May
3.	Jait	Middle of may to Middle of June
4.	Ashar	Middle of June to Middle of July
5.	Shrawan	Middle of July to Middle of August
6.	Bhado	Middle of August to Middle of September
7.	Kuwar	Middle of September to Middle of October
8.	Kartik	Middle of October to Middle of November
9.	Aghan	Middle of November to Middle of December
10.	Paosh	Middle of December to Middle of January
11.	Maha	Middle of January to Middle of February
12.	Phag	Middle of February to Middle of March

Table 2.15
Agricultural Calander

Sl No	*Crops*	*Local Terms*	*Sowing*	*Harvesting*
1.	Coarse Millet	Kutki	May-June	Sept-Oct
2.	Millet	Kodon	June-July	Sept-Oct
3.	Fine Millet	Samar	June-July	Oct-Nov
4.	Sorghum	Jowari	June-July	Nov-Dec
5.	Maize	Maka	May-June	Feb-Mar
6.	Pea	Rahar/Arhar	May-June	Feb-Mar
7.	Wheat	Pisi	Oct-Nov	Feb-Mar
8.	Gram	Chana	Oct-Nov	Mar-April
9.	Pea	Motar	Oct-Nov	Feb-Mar

Cultivation of Crops

Agricultural crops cultivation in the core zone villages of the Pachmarhi Biosphere Reserve, are of two types, viz. Cereals and Pulses. We have already discussed in the foregoing pages that not all crops are produced equally in all the villages under study. Crops are grown in different qualities in different villages as per the local conditions.

Agricultural lands are kept fallow for three months from March to May. Cattle go for grazing in those fields which do not have any fence. During summer they can not go to the forests due to forest fire. The cattle go to the fields which are without any protection and destroy the crops.

***Makka* (Maize)**

Every homestead has an attached garden (*bari* or *barha*), a fenced enclosure, about an acre in size, behind the house. In the *bari* mostly *makka* is grown. The grain which ripens first after the rainy season and thus helps in the trying time when the grain is exhausted and new harvest is still far off. Taking together all the seven villages under study, the total production of Maize is 17032 kilograms. All the villages produce a good quantity of Maize. Maximum *Makka* is produced at Rorighat (4070 kilogram), followed by Binora (3620 kilogram). The lowest amount of Maize is produced at Pattan (1262 kilogram). In Rorighat no Paddy is grown, since there is no land under Paddy cultivation. Being the only other major *Kharif* (monsoon period) crop, all the available land in *Kharif* season are allotted and utilized for Maize cultivation and as a result we get the huge production of Maize in Rorighat. No land in Rorighat is actually suitable for Paddy cultivation where retention of water is a must. Likewise in Pattan lands are so good and fertile that only a little amount of land is used for Maize, rest are used for more valuable crops like Wheat and Paddy.

In Rorighat lands in the hilly slopes are not ideal for Paddy but with a good monsoon the same land can produce good quality of Maize as a *Kharif* crop. In Binora all the *Bari* land is under *Makka* cultivation, it helps driving away the animals as it is attached to the house.

There are two varieties of Maize seeds one is hybrid, big in size and red in colour. They cultivate *Makka* and Paddy in two separate sections of the same plot at the same time. Preparation of soil for Maize starts in *Ashar*, and harvesting of *Makka* seeds mostly with the help of *nari*, i.e. seed funnel fitted to a plough. In Ghodanar *Barbati* is cultivated as a mixed crop, *Barbati* being a leguminous crop enriches the nitrogen content of the soil. *Makka* is sown only after preparing the land by *bhakar* at the onset of the monsoon, either by *nari* or by hand making a hole in the ground by using a stick. First they harvest *Makka*, then Paddy. Paddy is cultivated in the lower level of the field where water is available and *Makka* is cultivated at a higher level or on the slopes. In Badkachar several other crops are also make as a mixed crop. They sow *Jowari, Barbati, Jagni* (oil seeds) and *Urad* at the same type by broadcasting method. *Jagni* is considered a water consuming crop and hence it is sown wide apart from other crops in a big gap. Soyabean can also grow with *Makka*. First, they harvest *Makka*, then *Jowari* (2nd), then *Barbati* (3rd), then *Jagni* (4th) and at last they harvest *Urad* (5th). A raised platform called *Dhabua* is erected on the Maize field to drive away wild animals like wild Boar, Sambar, Monkey etc. At times they sow Maize with the help of digging stick by making a single hole and putting a single grain in it. By doing this the Maize plant will not fall in wind. In Rorighat and Kajri, there are three types of Maize seeds (1) *Bari Bhutta* (grown in *bari* land), (2) *Shankar Bhutta* (hybrid) and (3) White Maize called *Bhutti*. In Rorighat mainly *Bari Bhutta* is grown. They use both *bhakar* and plough in those fields only where grass is there and the soil is hard. Otherwise, in those fields where there is less grass and the soil is loose, they only use *bhakar* (leveler). To cultivate 1 acre of land 20 kilograms *Makka* seeds are required. If one has little amount of land he sows Maize with stick, if one has medium amount of land, he sows Maize by direct broadcasting method. If one has large amount of land he sows the seeds with the help of *nari* (seed funnel) attached to the plough. The cultivation of cereals and pulses in the same field is very useful in retaining the productivity of the land. Beside *Barbati* being a leguminous crop and it increase the nitrogenous contents of the soil. When there are three crops growing with Maize (*Makka*) in the same land as mixed crop, there is a calculation of ratio and proportion to be followed. If the total capacity of the land (area) can support sowing of 4 kilograms of Maize or *Makka* seed, where Maize is the principal crop. In mixed cropping there is one main or principal crop. The ratio/proportion would be 3 kilograms of *Makka* and 1 kilogram of *Jowari* or 2 kilograms of *Makka*, 1 kilogram of *Jowari* and 1 kilogram of *Arhar (Rahar)* pulses. The total seed(sowing) capacity of the land should not exceed the optimum quantity and at the same time, the ratio of seeds are to be maintained in mixed cropping.

People practise mixed cropping where there is a shortage of land. Mix farming is an attempt for optimum utilization of land resources. By mixed cropping all types of crops can be grown in the same land because they can not allot separate lands for different types of crops. But in Pattan *Makka* and *Tuar* are cultivated separately and they do not practise mixed cropping. Pattan has sufficient amount of land, as well as the fertile *dub- jamin* at the river side at their disposal and Pattan does not require to resort any mixed cropping as there is no paucity of land in Pattan. After the harvest the Maize grains are hung up in the field on wooden poles for sun drying or placed on a platform over the hearth fire for drying. Every day the women take some cobs, remove the grains by hand and dry them in the sun. After husking, they grind the grains to flour on the grinding stone. Mixed with water, the flour is then kneaded into flat cakes and roasted over a fire. Some times the women simply boil the Maize grains in water into a soft gruel *(Pej)*. Pouring of the water, they eat the gruel with buttermilk.

Paddy

Paddy is grown in *Khet* which is normally away from the home. They grow Paddy as *Kharif* (monsoon period) crop without the need for irrigation and transplantation. Paddy is sown in low-lying lands with black soil which retains the water. The Mawasi and Gond recognized four types of *Dhans* (Paddy).

1. *Kardana Dhan* – Black in colour.
2. *Batru Dhan* – White (*Safed*) in colour.
3. *Jubani Dhan* – Grows in juban land, i.e. low-lying fields with water retention capability during rains.
4. *Sarbati Dhan* – This is the best quality Paddy but it is not grown here.

Only three types of Paddy are actually grow here in Pachmarhi Biosphere Reserve area., of which *Batru Dhan* is the best in quality. In Pattan Paddy is sown in the month of *Ashar* after using *Bhakhar* and Plough by direct broadcasting method. After weeding for two times, harvesting is done in the month of *Kartik*. In Ghodanar and Kajri Paddy is sown with the help of *nari*. However, in a few places Paddy is sown by transplantation in *Ashar* and harvested in November after *deepavali*. For Paddy cultivation they use cow and goat dung as organic manure. They keep the dung on the field itself in a heap, after 3 to 4 months they mix the manure in the field. They never dig a pit to prepare compost manure. Maximum quantity of Paddy is produced in Pattan (7950 Kilogram), followed by Ghodanar (1012 Kilogram) and Kajri (850 Kilogram). Dhana, Binora and Badkachar produced 455 Kilograms, 400 Kilograms and 251 Kilograms of Paddy, respectively. Rorighat does not produce any Paddy. Pattan has maximum quantity of land under *Kharif* cultivation and a good proportion of it is used for Paddy cultivation. Ghodanar, Kajri and Dhana have got low-lying *juban* land which is ideal for Paddy cultivation. Rorighat does not have any low-lying land suitable for Paddy cultivation.

Kodon / Kutki

Kodon (Paspalum scrobiculatum) and *Kutki* (*Pancum* mileaceum) are the two types of coarse millets cultivated by the Mawasi and Gond tribes living in the Pachmarhi Biosphere Reserve core area. These coarse grains can grow with very less amount of water and in harsh soil conditions. Samar is another variety of coarse millet, similar to *Kodon*, cultivated in Pattan village only. *Kutki* is the typical crop of the shifting cultivation, which the Mawasi (Korku) are said to have been cultivated in the past. *Kutki* is a small kind of millet, which grows on all soils but preferably on red soil. After the harvest, the grain is brought to the threshing floor where an open space has previously been plastered with a mixture of mud and cow dung. A wooden post is fixed in the centre of it. The *Kutki* is then spread over the threshing floor and four or five bullocks are tied together with a rope and the end of which is tied to the post. The animals are driven around the post until all the grains have been threshed out. The winnowing is later done by pouring the grain with the husks from a basket held high against the wind, which blows off the calf. The grain is then stored in the big containers woven with bamboo splints and plastered with a mixture of mud and cow dung. *Kutki*, after husking, is boiled into a gruel and eaten with some vegetables.

Rorighat tops all the villages in the matter of *Kutki* production as well as the land under *Kutki* cultivation. Rorighat grows 7880 Kilograms of *Kutki.* Binora and Pattan grows 400 Kilograms and 500 Kilograms of *Kutk*i, respectively. Rest of the villages produce hardly any *Kutki. Kodon* (1250 Kilogram) produced only at Binora. Samar (4405 Kilograms) is produced only at Pattan. Due to the scarcity of water *Kutki* is grown more at Rorighat. Bad terrain and stony soil in sloping field promotes *Kutki* cultivation in Binora. However, *Kodon* at Binora is grown in good plain land in rotation with Wheat in the monsoon to maintain and regain the fertility and productivity of the land. In Rorighat the people possess some vested revenue land called *rajas* jungle which is not forest land but located just outside the periphery of the village, actually situated in between the forest *munara roman* and village homestead lands and fields. In these *rajas* jungle lantana or *juranta* grows profusely and spontaneously in monsoon. The people of Rorighat produce *Kutki* in these land by a unique method of slash and burn (controlled) method of cultivation. They cut the plants, dried in sun and burn. The ash acts as good fertilizer. After the first year of cultivation, the land remains fallow for two consecutive years, i.e. 2nd and 3rd year, then again in the fourth year they cultivate *Kutki* in the same land. Here we can see that though *dahiya* (slash and burn) cultivation has been stopped, it is followed here in Rorighat in a different way. Binora also produce *Kodon* and *Kutki* in alternate years. There are two varieties of *Kutki* seeds. *Bari arma*, which is grown in Rorighat in slash and burn method and *piria Kutki*, which is grown in Binora by *bhake*r and a pair of bullock in normal cultivation, cultivation of *Kodon roman Kutki* by using plough and *bhakar* as practised at Binora is quite unique. In Binora the lands under *Kodon/Kutki* cultivation are kept fallow for one year after every three years of consecutive cultivation. Even they cultivate *Kodon* and *Kutki* alternately in the same land.

Preparation of the field for *Kutki* cultivation starts in the month of *Shravan* by field preparation with plough and *bhakar*. They sow the seeds in *Shravan* itself by direct broadcasting method. Harvesting of field *Kutki* is done in the month of *Aghan*. *Piria, Arma* and *Rosary* are the three main types of *Kutki* seeds. *Piria* type of Kutki can only grow in the hill slopes where there is no water. *Aarma* and *Rosari* type of *Kutki* can grow only in the field (*Khet*) by using plough and *bhakar*. In Binora village for *Kodon* cultivation, unlike *Kutki*, they do not use plough. They till the field with *bhakar* only in the month of *Ashar*. To cultivate *Kutki* in 1 acre of land 40 Pai (i.e. 120 Kilogram) seeds are required, the production output will be 800 Kilograms. To cultivate *Kodon* in 1 acre of land 8 Pai (i.e. 24 Kilograms) seeds are required and the production output will be 1000 kilograms. *Kutki* is harvested around *Deepavali*. *Kodon* is grown as mixed crop with *Makka* but *Kutki* is grown separately.

Method and time for the cultivation of *Kutki* vary from place to place, and even in the same village there could be some variation. *kutki* which is grown in agricultural field is also called *Khetai Kutki* and the one which is grown in slash and burn method by burning lantana plant (in Rorighat) is called *Bewar Kutki*. *Bewar Kutki* plants are bigger in size than the *Khetai Kutki* but labour input is more in *Bewar Kutki*. Again as per information there could also be three more varieties of *Kutki* seeds, e.g. (1) *Piria Kutki*, (2) *Kali Kutki* and (3) *Dhar Kutki*. *Piria Kutki* is sown with *Rahar (Arhar)* in the month of *Ashar*. *Arhar* comes out first in the month of *Kartik*, after that *Kutki* comes. In the villages of Pachmarhi Biosphere Reserve core area *Kutki* is cultivated more than *Kodon* because for *Kodon* cultivation bullock is needed but cultivation of *Kutki* can be done even without bullocks. Rorighat has little or no bullocks so they cultivate more *Kutki* and little *Kodon*.

Wheat

Wheat is the principal winter (*rabi*) crop in the Pachmarhi Biosphere Reserve core area villages. Both the Mawasi (Korku) and the Gond give importance to the cultivation of Wheat is evident from the fact that it is the only crop for which they use chemical fertilizer like DAP and Urea. Wheat is sown in November or December and harvested in February or March. It grows in black soil in low-lying areas where moisture is retained long after the rainy season. Wheat can be of two kinds, a red kind which has to be watered and a white kind, which grows without irrigation. As soon as the monsoon crops have been reaped, they plough the plot suitable for Wheat cultivation and cleanse it from the stubble of the earlier crop with the harrow. The sowing must begin when the ground is still moist, usually in October. The sowing of the Wheat cannot be done with the three-pronged tifan, for the drills would soon get blocked in the wet soil. The farmers fix the bamboo tube through which the Wheat seeds fall down behind the ploughshare. While he draws furrows in the ground with the plough, a woman pours the seed into the seed drill. The seeds fall immediately after the ploughshare, has opened the ground and go into the soil, which at this depth is still moist. The next furrow which is just six inches distance closes the furrow hole which has received the seed a short while ago. Thus, the wheat seed is all embedded in moist soil and can grow without any rain. The wheat is reaped in February or March. It is cut with sticks, tied into sheaves and brought to the threshing floor where it is threshed and winnowed like other grains. Wheat is not sold in the market unless it is absolutely necessary. The Mawasi and the Gond prefer wheat bread to any other bread. Wheat cultivation in the irrigated land with the use of pump is increasing day by day in the Pachmarhi Biosphere Reserve core zone villages.

Maximum quantity of wheat is produced at Kajri (15880 Kilograms), followed by Pattaṇ (12500 Kilograms) and Rorighat (4700 Kilograms). Production of Wheat at Kajri is maximum amongst the seven villages under study. The lands under the wheat cultivation at Kajari are very fertile because, due to their preoccupation in the *Nagduari Mela* during monsoon most of the low-lying good lands are used for *Mela*. Since these lands remain fallow in the *Kharif* season their fertility comes back to an optimum level to give a bumper crop in the next *rabi* season in the form of wheat. The income generated from the *Mela* is also spent to buy chemical fertilizer, which helps in the increased amount of production of wheat at Kajri. The village Kajri used highest quantity of chemical fertilizer. In total 2190 Kilograms, of which DAP is 1000 Kilograms and Urea is 1130 Kilograms. In Pattan wheat cultivated in done in *the dub jamin*. Silt deposition increased the fertility of the fields which in turn caused increased production of wheat. Similarly, Rorighat also uses 727 Kilograms of chemical fertilizers in wheat cultivation, which is instrumental to its high wheat production.

Wheat (*Pishi*) is normally sown in *Aghan* and harvested in *Chait*. There are various types of wheat (*Pishi*). *Dhania Pishi* looks like *Dhan* or Paddy. It also requires more water like the Paddy to grow. *Dablu Pishi* is white in colour and is mostly grown in the area. *Lokman Pishi* is costly and it produces good quality flour for soft *roti*. *Sarbati Pishi* is also like the *Lokman* but more costly. There are some more varieties of wheat that are also available. Such as *Naramadachar* and *Bhakarwali Pishi* which is the local red variety of wheat. DAP and Urea are used in wheat cultivation in the core area villages of the Pachmarhi Biosphere Reserve. DAP is mixed with the wheat seeds and the grain in sown with the help of seed funnel. After watering for two or three times when the wheat crop grows on the ground, urea is sprinkled on it. In the same plot of land after harvesting of *Makka* and *Barbati* they sow wheat and *Matar* separately. Because wheat and *Matar* (Peas) can not grow in the same field together. *Matar* is sown in *Paush* (December) and it is ready by the net month of *Maha*. When *Chana* (gram) and wheat are cultivated in same ground, *Chana* grows in the side and wheat grows in the centre of the field. In Pattan, however, they sow Wheat after *Chana* is harvested or they grow *Chana* and wheat in the same land on a fifty-fifty basis.

Social Organization of Production

A pair of bullocks is hired @Rs.50/- per day and a man gets Rs.70/- per day for agricultural work. But when both are taken together as a package, man and bullocks, the rate would be Rs.100/- per day from 8.00 a.m. to 12.00 noon, after a break, again from 2.00 p.m. to 4.00 p.m. In Badkachar village people use to get Rs.20/- per day for agricultural labour work within the village for weeding of crops like wheat, *Makka,* etc. In Kajri when daily wage agricultural labours are engaged from outside the village, the rate is Rs.60/- per day but when villagers work for the persons of the same village they get only Rs.20/- per day. In Badkachar village people use to get Rs.20/- per day for agricultural labour work within the village for weeding the crops like *Makka*, wheat etc. In Kajri when daily wage agricultural labour is engaged from outside the village the rate is Rs.60/- per day but when the villagers work for the persons of the same village they are required for pay only Rs.20/-. Harvesting of wheat and some other cereals have some other mode of compensation. They get 1 *pura* of wheat after harvesting

19 *puras* of wheat i.e. in the ratio of 19:1. *Pura* is a bundle of Wheat tied together for transportation from field. For every 19 *puras* of harvested wheat the workers get 1 *pura*(i.e 7 kg) of wheat. In Binora we come across a unique system of barter and exchange involving man and bullocks. If someone takes a bullock from a person free of charge, he then himself has to go for work in that person's field in lieu of cash payment. A pair of bullock works in the field from 8.00 a.m. to 3.00 p.m., but in exchange, the man has to work from 8.00 a.m. to 5.00 p.m.

Cultivation of Other Crops

Besides the major crops, mentioned here, there are a lots of other crops, which are being cultivated in the Pachmarhi Biosphere Reserve core area villages under study. These are discussed in the following lines –

Parasbeans and Green Peas are import crops. Green peas grow with wheat and *Parasbean* with *Makka*. They cultivate *Parasbean* with Maize and support the creeper with a pole so that it climbs the Maize plants. *Barbati* is of two types one is creeper (*bela*) which climbs the Maize plants, other variety is a smaller one. The shrubs do not climb the Maize plants. In Kajri there are three variety of *Barbaties*, red, pink and white. White one can grow in less water but the other two require some water. Smaller *Barbati* is called *Rajma*. In Pattan another local variety of Millet (smaller) is called *Sama*. For *Sama* no weeding is required. In Pattan they cultivate *Sama* and *Makka* separately because for *Makka* weeding is required but not for *Sama*.

Weight and Measurement

The smallest unit of measurement is called *Paua*, 3 *Paua* is 1 *pai* (3½ Kilogram), 24 *pai* is 1 *maund* (32 Kilograms). A 25-Kilogram bag (*bori*) holds 22 *pai* and 1 *bora* contains 100 *pai*.

Management of Forest Resources

Each village of the Pachmarhi Biosphere Reserve core zone has its definite and demarcated forest area for exploitation, which is a radius of 5 Kilometers. In this area only cattle (Cow or Goat) of that village can graze. Cattle of other villages are not permitted in this area. Since there is no such restriction on the movement of humans, men and women can roam around for collection of firewood and other forest produces. For collection people go out of their home as early as 4.00 a.m. in the morning to the forest of other villages' vicinity also. Collection parties are constituted in groups with persons of same sex. Males and females form separate groups. Only the husbands and wives can go together in the forest.

The villagers do not buy fodder from the market. They take their cattle for grazing in the forest. *Khandi, Moya* ad *Sukra* grasses are mainly for the cows. Goats eat only leaves but cows eat grass which creates problem of finding new pasture for the cows. Goats and cows can graze in the same place since their eating habits are different. Goat goes to higher land and eats leaf whereas cow grazes in lower land in the meadows and eats grass. Cattle herders are called *Charhawa*. Charges for cattle herding is Rs.25/- per month for one cow or one goat. Herding inside the National Park is not permitted so the herders take their cattle only to the sanctuaries. In case a cow is eaten by tiger, forest department pays compensation of Rs.2500/- to the owner, however there is no such thing for the goats.

Rambuhari (broom) and honey are the main forest produce of Kajri. Beside *Rahar*, i.e. extraction of Resin (*dhoop*) from *sal* and *sarai* trees, collection of *achar (chirongee* – buchanania latifolia) and mushroom (*bhomori*) are also present. *Rambuhari* broom (*jharu)* is sold @Rs.5/- to Rs.8/- per piece. *Achar* is sold @Rs.40/- per Kilogram, however they sell it after making *chironjee* at a much higher price. *Mahul* leaf is sold @Rs.3/- to Rs.8/- per Kilogram in the Pachmarhi market. Broom of Chinda grass is sold Rs.4/- per piece. But now a days collection of *Chinda* grass is not permitted. There are two types of Mushrooms, white and black. The white variety, which grows on soil, is used as eatable vegetables. It is very hot and body becomes warm after eating. The black one which grows on wood is poisonous and it is not eaten as food. In Kajri, they do not collect *Mahul* leaf because selling of *Patta* daily at Pachmarhi market is not possible from here. They collect firewood at Kajri also for their own consumption, which can not be transported and sold at Pachmarhi which is far away.

Mango is profusely collected as a forest produce in Rorighat. The village people know the definite names of the forests in their vicinity. In which direction and at what distance the forest is located. What species of plants and trees grow in which forest. Even the cattle have an idea where to get grass or leaf and they tend to go at that direction only. In Rorighat *achar* is available less and Mango more. They do not sell *achar* in this village. One *bhoja* (i.e. 70 to 75 Mangoes) of ripe Mango is sold @Rs.25/- per *bhoja*. Green *(Kacha)* mango is sold @Rs.150/- per *bori*. There are 150 to 200 mangoes in a *bori*. If all the mangoes in the *bori* (bag) are from the same tree, then the price will be more. Green mango is more costly because it can be transported for business purpose for making of pickles in the industry. The ripe mango can only be sold locally for consumption at local level and cannot be transported. Whether a mangoes is sweet or not, is ascertained by the fact that it is eaten by the birds or not. Because they know that birds eat only sweet mangoes. In Binora the most sought after forest produce is *Mahua* (bassia latifotiatrae). *Achar* is also available at Binora. Demand of *Mahua* at Binora is for commercial purpose only. They make and sell *Mahua* liquor in the *Sivaratri Mahadeo Mela* (held in 8th of March) to the pilgrims. The distillation process is conducted in the Denua riverbed. A bottle of *Mahua* liquor (½ Litres) is sold at the fair (*Mela*) at Rs.10/- per bottle. In Binora, on an average, each family earns Rs.1000/- to Rs.1500/- during the *Mahadeo Mela*. In Binora village they do not sell *Mahua* flowers which is to be sold at *Mela*. The *Mahua* flowers are kept in a covered container, little yeast is added too, after some days the smell of alcohol comes and the flowers become fermented. They boil it on low fire in the Denua river bed. The steam goes out through a pipe and enters into another container partially submerged in the cold flowing water for distillation.

Pattan village does not have much of forest collection. The land available at Patta for cultivation, including the *dub jamin*, is plenty. The forest near Pattan is infested with wild dogs (*son kutta*) due to this their domesticated dogs avoid to accompany them to the forests and they are also reluctant to go to forests without their dogs.

People's participation in forest management is organized through local *Samities.*There are Eco-Vikas Samiti in Satpura National Park and Van-Suraksha Samiti in Bori and Pachmarhi Sanctuaries. These Van Samities have 12 members, including the forest beat guard as being the Ex-Office Member Secretary. They observe the behaviours of the people in relaticn to the management of forest environment. They see to it that the exploitation of forests is according to the law. They employ people for forest work and pay Rs.70/- per day out of which Rs.10/- is deducted by the Samiti to create a cash reserve fund for welfare purpose. People take loan from it @ 2% interest to meet their needs. The capital reserve of the Eco-Vikash Samity of Ghodanar is Rs.50,000/-.

The best season for forest collection is from February to May, i.e. during summer, collection time for Mango is *Jet* and *Baishak. Mahua* is available from *Phag* to next two months, i.e. upto *Chait* and *Baishak. Rambuhari* (broom) is collected in the month of *Paush* and *Magh*. Honey is collected from *Baishak to Jet/Ashar.*

Bamboo is an important and very useful forest product. For making strips *tatta* or *karler* bamboo is used. In general ordinary bamboos are named differently as per its age or how old is the bamboo plant. *Mahta* or two years old bamboo is used for making of *Supa* and *Tokri*, *Dehla* and three years old bamboo is normally used as fence and outer walls of the house. *Sukha* or five years old bamboo is the mature one, it is used for house construction. The people understand the precious value of the bamboo and that it takes a long time to regenerate a bamboo groove. The people never sell the bamboo in market but only use it for self-requirement.

Honey collection is another important economic pursuit from forest. Honey is generally got from two kinds of honey bees, a large variety called *agia* and a small one called *jharia*. The large variety clusters mainly on *Koha* trees (terminalia arjuna) but the *jharia* variety gives the better honey. There is still another kind of bee called *bhaunra* which builds the hives on rocks. Rorighat village is famous for its expertise in honey collection. Maximum honey collection is done in the month of Baishak because flowering of the trees like Mango, *Jamun, Mahua, Bhandari, Parsa (Palash), Bertha*, *Dhaman Butthi* etc., take place only two to three months before *Baishak*. So *Baishak* and *Jeth* are the two months when honey is collected most, once monsoon comes in Ashar the honey bees themselves eat the honey. During *Kuwar* (September/ October) month there is a second season of honey collection. This honey is not of good quality and there is no guarantee for it, the price is also much less for this honey. One bottle of such *Kuwari* honey is sold @Rs. 30/- to Rs. 40/- per bottle only. They usually collect honey during the dark period of the moon when honey is stocked more. During dark period of moon bees bite less and more honey is fetched, at a time upto 20 to 25 Kilograms. Honey is always collected at night because during day the bees never allow to get near the rock or tree where the honey hive (*bhonr*) is located. They tie a bundle of grass around the top of a bamboo, climb up the tree and set fire to the grass. They thrust the burning torch into the beehive. The light blinds the bees so that they fail to see the man and do not attack him. They also make ropes from the bark of *mahul* tree and build a saddle. It is then hanged from the rock (cliff) side or tree, now sitting on that hanging seat a man cuts the hive and brings it down gently with rope. They separate the honey from the wax by applying pressure. A clean cloth is used to separate the honey. They leave a small portion of the beehive intact *in situ* so that bees can again make the hive at the same place next year.

Medicinal Plants

Medicinal plants are extensively used in the villages within the Satpura National Park of the Pachmarhi Biosphere Reserve for the treatment of diseases and ailments. However, the identification of the plants are only restricted to the specialists, and not to the mass. These knowledgeable few not only knew the useful plants and their parts to be used but also the preparation and formulation of the doses.

Here under are given the remedies for some common ailments –

Fever, Cold and Cough and Malaria etc.

1. *Tulsi* (ocimum sanctum) leaves can cure common cold, cough, sore throat and ulcers.

2. Kadua (*Bitter*) Chirat (*Andographis Paniculata*) herbs is sour in taste. A paste is made of its leaves and then by using a sieve the juice is separated and given to drink. It relieves the fever.
3. Mita Chirt (*Swertia Chirata*) plants are also used in the treatment of fever.
4. Rakat bera leaves are boiled in water, it cures fever, body pain and cough.
5. Harrha (*Terminalia Chebula*) fruits are fried and eaten with a mixture of tulsi and honey. It is believed to have given relief in cold and cough.
6. Roots of bhuth kesar (*Crocus Sativus*) herbs are burned and the smoke when inhaled by the children cures throat infections in children.
7. Bihi (*Guava*) is fried and taken for running nose.
8. For acute cough and cold and fever- Bark of Mango (*Magnifera Indica*), bark of harra (*Terminalia Chebula*), leaves of ber (*Zizybus Jujube*), roots of daru haldi (*Berberis Asiatica*), bark of jamun (*Lugenia Jambolana*), bark of behra (*Terminalia Belerica*) and bark of mahua (*Bassia Talifolia*) are mixed together and are kept in earthen pot for 2 to 3 days. A paste is made with it and taken as a good remedy for the ailment.
9. Tulsi (*Ocimum Sanctum*) – The worshippers of *Narayana* must worship tulsi which is believed as the incarnation of goddess *Lakshmi*. This is very useful plant – leaves of which can cure common cold, sore throat and ulcers.

Diarrhea, Stomach upset and Indigestion

1. Jugni (a type of oil seed) and Mahua seeds are made into a paste, stirring by a wooden stick, and given for stomach upset. It tastes like Bournvita.
2. Bark of simul tree (bambax malabaricum) i.e. cotton, is made into a paste and is taken mixed with sugar.
3. Bihi (guava) leaf paste is taken as digestive.
4. Sal (soree robusta) tree's rasin is used as astringent and in weak digestion.

05. Ber (sizyphys maritiana) fruit is used for diarohea and dysentery.

Skin Diseases

1. Van haldi (*Curcuma Aromatia*)- root of the shrub is used for skin diseases.
2. Haldi (*Curcuma Longa*) – A paste of turmeric and oil is applied all over the body of young women as a skin conditioner, antiseptic agent and germicide.
3. Neem (*Azardirachita Indica*) – All the parts of the tree is used in the treatment of ulcers and skin diseases.
4. Van-vamtilla (*Vicoa Auriculata*) leaves are made into paste and given in diarrohea.

Plants and Treatments used for some other diseases and ailments

1. Twigs of jamun (*Syzygium*) tree is used as *dantan*.
2. For vomiting – peacock's feathers are burnt and mixed with honey. The paste is taken two to three times in a day. This will stop the vomiting.
3. For skin diseases – Roots of karwa chirat herbs (*Swertia Chirata*) is boiled in water. Bath is taken in that water, it cures skin diseases itching, ring worms and boils, etc.
4. Cataract of eye – Mahua (*Madhuka Indica*) bark, mango bark, harra bark (*Terminalia Chebula*), behra bark (*Terminalia Belerica*), daru haldi root (*Berberis Asiatica*) and rakat bera leaf are boiled in a vessel. A thick compound is prepared by stirring the water. The thick compound is kept in a red earthen pot (*Matka*) for three to four days. Eyes are washed with it for the treatment of cataract.
5. Foot Ulcers – They apply hot *harra* fruit in ulcers of the foot. It cures the wound soon.
6. Bone Fractures – Harjuri (*Vitis Quadrangularis*) creeper is used to treat fracture. A paste is prepared by grinding the leaves and then it is applied over the affected parts. They also eat the paste for speedy recovery.

7. Child feeding – For the increase of breast milk in women. A paste made of the mixture of bhojraj and tejraj (*Xanthium Alatum*) root is given to the mother for the increase of her milk.
8. Dental pain – They boil the bark of ber (*Ziziphus Mauritiana*) tree in water and wash mouth in that water. This will give relief from dental pain.
9. Ear diseases – To get relief from ear pain they use the milk of *rattan joth* stem.
10. Burn – The milk which comes out from *gamar* herb's (*Gymnema Sylenestie*) stem is used on the burn wound for relief.
11. Twisting of ankle – Roots of daruhardi (*Beberis Asiatica*) herbs is taken with tea.
12. Blood purification – Roots of daruhardi is taken with tea.
13. Increase of hemoglobin in blood – Roots and fruits of Tejraj herbs and fruits of Bhojraj herbs are mixed together and fed with milk.
14. Chicken pox – Chicken pox is called Chachak mata. Extracts of neem leaf is given for relief.
15. Cataract (mature and old one) – Extracts of tubers from chamali herbs is taken, a paste of the same is applied on eyes.
16. Mental problem – Tubers of bhakkhand creeper is sundried and made into a powder is then given to the patient mixed with elixer of ajjwain (*Acoaptycotis*).
17. Arjun (*Terminalia Arjuna*) – Bark of the tree is used for heart ailments.
18. Vanhaldi (*Curcuma Aromatica*) – Roots of the shrub is used for skin diseases.
19. Haldi (*Curcuma Longa*) – A paste of turmeric and oil is applied all over the body of the young women as a skin conditioner, antiseptic agent and germicide.
20. Munga (*Moringa Olifera*) – Bark of the tree in paste form is used as poison for arrow heads.
21. Bach (*Acorus Calamus*) – Root of the herb is used for vitality, blood purifier and cold.
22. Safed musli (*Chlorophytum Tuberosum*) – This is called Indian Viagra. It is used for vigour and vitality in human males.

23. Aonla (*Emblica Officinalis*) – Fruit of the tree is used as laxative and asthama.
24. Jangali piyaj (*Onion – Urginea Indica*) – Bulb of the herb is used for paralysis.
25. Kalihari (*Gloriosa Superba*) – Plant itself is poisonous but root is considered as antidote in cobra poisoning.
26. Tejraj – Root of the plant cures impotency in men.
27. During monsoon, wearing of wet cloths caused fungal infection. As a treatment mushroom is dried in sun or over an oven. The infected person is made to lie down on a bed made of ropes under the bed boiling water is kept and the dried mushroom is put into the boiling water. The vapour comes out from it would cure the fungal infection in the affected portion of the body.
28. To cure pneumonia in children, three earthworms are taken out from *juban* soil (Paddy field), one earthworm is cut into two halves. Half piece of the earthworm is put back into the soil. The rest two and half pieces of worm are sundried and a paste is made. This paste is fed to the child to cure acute pneumonia. This is also given to the mother of newborn to increase the breast milk.
29. Spines of porcupine are burnt and then it is mixed with oil and applied to old wound for healing.
30. Flowers of bhuth kesar herbs, which grow on stone, are kept with food grains for preservation while storing. Neem leaves are also used for storing of grains.

Summary and Observation

The human dimension of the Biosphere Reserves make them unique conservation units which are different from other protected areas like Tiger Reserves, National Parks and Sanctuaries. In Biosphere Reserve interaction with local communities is essential to keep intact the ecological knowledge and cultural values of the area by conserving natural resources with their sustainable use. Hence, the use of indigenous knowledge (a set of perception, experience and behaviours) that guides local people in the use and management of natural resources effectively within the Biosphere Reserve is an important issue.

With this objective in mind all the seven village settlements situated within the core area of the Pachmarhi Biosphere Reserve have been studied. The entire core area of the Pachmarhi Biosphere Reserve is coinciding with the geographical unit of the Satpura National Park. The situation in each of the six villages is different from others, so are their land use pattern, cropping patterns and agricultural operations. The variations observed are due to adaptation to the local ecological conditions. The local people managed their natural resources sustainably with the help of their indigenous knowledge. This has been studied and village wise differences are noted. It is observed from the tables that some villages grow more Kodon or Kutki and some grow more wheat. The reasons behind it has been investigated and clarified. Some use more fertilizer than the others. Thus, Rorighat produces more Kutki than others; Paltan produces more wheat and Chana than others. Kajri uses more fertilizers than the other villages. It is also shown how religion and ritual shape the village economy, as in the case of Kajri and Binora, due to the celebration of *Nagpanchami* fair by the pilgrimage of *Nagduari* Cave at Rorighat and Mahadeo fair in *Sivaratri* at Binora. Similarly, collection of forest produce in a particular village depends on its availability in the surrounding forests as well as the facility of its marketing. *Mahua* collection is more in Binora due to its scope of selling during the fair. Honey collection is more in Rorighat because beside the help of nature, it has a good market in Pachmarhi town. Peoples knowledge in the selection of seeds and collection of forest produces has been discussed in the foregoing pages. It may be mentioned here that the knowledge of medicinal plants and its usefulness are legendary in the villages of Pachmarhi Biosphere Reserve. An attempt has been made to mention the specific plants with their local and botanical names.

Reference

1. *Pachmarhi Biosphere Reserve*.Vols One and Two, Environmental Planning and Co-ordination Organisation, (EPCO). Bhopal, 2001.

3

Environmental Orientation in the Context of Traditional Tourism in a Biosphere Reserve

D.K. Samanta

During the past few decades, tourism and environment are the two important topics which have attracted the attention of the anthropologist. Protected areas typically promote nature tourism i.e. travel to less spoiled natural places where people can experience and enjoy the serenity of nature. Of late eco-tourism has been promoted as an alternative, low impact form of tourism to such areas rich in natural resources. It is viewed as a combination of conservation and development tool because it provides conservation as well as economic benefits. Ecotourism appears to be the fastest growing sector of the tourism industry as it has an educative and respect for indigenous cultural elements. At the same time it has raised the issue of the possible degeneration of natural resource. Such protected natural resource areas may be associated with sacred places and having a long tradition of religious tourism. The human element in the form of locales is an important factor in this arena. This changing orientation to tourism, environment and people has hardly paid any attention to people's orientation to these aspects. This conjunction is our concern in this study. To understand the intricacies of this process the study of a Biosphere Reserve became imperative. Pachmarhi Bio-sphere Reserve provided an ideal situation for conducting such a study.

In nature tourism there are tourist and nature. It is a travel to unspoiled natural places where people can experience and enjoy nature. Whereas ecotourism includes socio-cultural elements, seeing and interacting with people whose customs and rituals seems exotic and attractive which deviates from the cultural experience of the nature-tourists. It is a visit to places of natural beauty that conserves the environment and sustains the well being of the local people. In contrast religious tourism involves people i.e. Pilgrims from other areas to fulfill their psychic need and vows. The local people get involved in the process providing material and services. There is a symbolic attachment with the deity and the people around and vice versa guided by age old traditions. A pilgrim is over concerned about the sacredness of the environment around the place of pilgrimage. And on the other hand, the host population's support to the pilgrims and help extended is guided by the concept of helping a pious person or a person attempting a pilgrimage.

The local people require to orient themselves to such large scale influx of pilgrims in their sleepy environment during specific period in a year. It is marked by some auspicious occasion indent in tradition passed on orally from one generation to the next. They are to accommodate the itinerary of the occasion in their daily routine of works.

Protected areas promise nature tourism turned eco-tourism which follows favourable and unfavourable seasons. The favourable seasons involve vibrant activities to eco-tourists' destination. This too entails orientation of the local population to such process. Ecotourism can be a well neigh alternative as a low impact form of tourism to natural areas as compared to nature tourism. It is a concept gaining ground on a faster pace.

A natural area which is already a destination for religious tourism marching ahead for nature tourism and ecotourism in recent years may be an ideal area for understanding the dynamics of orientation to such changing process. The present effort is a primary step in that direction which has been attempted by highlighting the process of religious tourism in a natural surrounding. It was imperative to select a natural area where both the process can be observed in phases. Pachmarhi Biosphere Reserve provided the ideal situation for the purpose. The dramatic beauty of the natural landscope, abundance and diversity of flora and fauna, rich archaeological resources, sacred hills and an abode of local Shiva combine to make the area scientific, ecological and its cultural resources exceptional. The concept of conservation has its origin in this bio-reserve which dates back to nineteenth century under British domination. Its diversity and richness has earned the name of Botanists' paradise. *Pachmarhi* has in its store several cave paintings of great archaeological significance depicting various subjects like hunting, faunal varieties, etc. The *Pachmari* is rich in forest resources. Sal and teak, two most valued timbers in market are in abundance in the Bio-reserve. The entire forest can be broadly classified into moist deciduous, dry deciduous and central Indian sub-tropical hill forests.

The important caves having old rock paintings include Mahadeo, Catacomb, Jatashanker, Pandav caves and Mandiadeo which have archaeological importance. The hills around *Pachmari* are considered as sacred being the abode of local Shiva or Mahadeo. The *Handi-khoh*, a deep gorge is believed to be the retreat of a monstrous serpent. *Jambu-Dweep* is also a gorge cut into soft rock at the bottom of which is a cavern in the rock resembling a giant lock of matted hair called Jatashankar. *Chauragarh* which is the second highest point in *Pachmarhi* ranges only next to *Dhupgarh* houses the temple of Mahadev. All of these have added sacredness to *Pachmarhi. Nagpanchami* in July-August (Sravan) and Mahashivratri (March) are the two important Hindu festivals when lakhs of pilgrims from the plains visit these temples.

The study provides the details of religious tourism on the eve of Mahashivratri which is in vogue since immemorial times. Part one provides the features of Pachmari Bio-Reserve and historical backdrop. Part two describes the traditional route and the cluster of seven villages remain directly involved in the yatra. Part three deals with the impact of this yatra on the people of this village cluster. Part four discusses the role of the tribal people MawasiThakur/Korku and Gond. Part five depicts the process of orientation operative in the area. The study on which this paper is based was originally carried out for the purpose of preparing the case study of eco-tourism in the Pachmarhi Biosphere Reserve and to understand the interplay of dynamics of orientation and sustainable development.

The Satpura hill range is one of the major geographical features of India. *Dhupgarh* peak located in *Pachmarhi* hill station in Hoshangabad district of Madhya Pradesh is the highest point in the range. The Pachmarhi Biosphere lies between latitude 20° 10 to 22 ° to 22° 50 N and longitude 77° 45 to 78° 56 E constituting parts of Hoshangabad, Betul and Chhindwara districts. It is a multiple land use zone in which core and buffer zones are well delineated.

It is the world body UNESCO which initiated the concept of Biosphere Reserve in 1973-74 under its Man and Biosphere programme with a view to conserve the depleting and endangered gene pool and thereby maintain the ecological balance. India being one of the twelve mega biodiversity countries of the world took initiative to participate in this movement in 1986. It represents two realms and three basic biomes and has ten bio-geographical zones. As per this programme representative parts of natural and cultural landscapes extending our large areas of terrestrial or marine eco-system or a combination of both are designated as Biosphere Reserve. There are in all twelve Biosphere Reserves in the country. The Pachmarhi Biosphere Reserve in central Indian state of Madhya Pradesh was declared as Biosphere Reserve in March 1999. It has the characters of a Bio-Reserve like representative terrestrial environment, recognized for its value in conservation and providing scientific knowledge, skills and human values to support sustainable development. The Bio-Reserve also includes representative examples of natural and minimally disturbed eco-systems e.g. endemic centers, genetic richness, suitable for experimental manipulation harmonious landscapes due to the continuation of traditional pattern of land use, etc.

The total area of Pachmarhi Biosphere is 4987.38 sq.km, and comprises three wild life conservation units Bori Sanctuary (485.72 sq.km.) Satpura National Park (524.37 sq.km.) and Pachmarhi Sanctuary (417.78 sq.km.). It envelops 3213.32 sq.km, forest area, 1364.56sq.km, farm land, 122.01sq.km, waste land and 204.3 sq.km under water bodies. An area of 2730.48 sq.km accounts for closed forest while open and degraded forest comprise 264.49 and 136.03 sq.km. respectively. The area comprising of Satpura National Park, Bori and Pachmarhi sanctuaries has been notified as project Tiger in August, 2000. Satpura National Park comprises the core zone and the remaining area of 4462.93 sq.km, surrounding the core zone serves as buffer zone. The spread of the core zone i.e. Satpura National Park is 524.37 sq.km. In the north it is bounded by road from Tawa left bank canal north of Pathorota to Sohagpur, Shobhapur Piparya upto river Dudhi. River Dudhi form the eastern limit and also marks the district boundary between Hoshangabad and Narsingpur and Chhindwara and Narsingpur. The metal road from Jamai to Damu and along the boundary between Chhindwaraand Betul districts, towards west along pucca road via Hirapur stretching to Shetpur forms the southern boundary of the Bio-Reserve. The Nagpur-Bhopal road form the western limit of the Pachmarhi Biosphere Reserve.

The attitude varies from 320 to 1352 m above msl. It is a hilly undulating terrain which runs from east to west, precipitous at places, having deep gorges around Mahadev hills in Pachmarhi plateau. The Pachmarhi plateau is almost at the centre of the Satpura hill ranges running east-west. Dhoopgarh is the highest point (1352 m) in the range wherefrom river Denwa rises takes eastern and then northern course to meet river Dudhi which in its turn drains into Narmada. The Pachmarhi hills have steep slopes to the north and in the south it is straight exposing the base rocks. Jambudweep, Handikho, Mahadev, Chauragarh and Bee Pall are some of the conspicuous hilly areas. The Aschcan phyllites and Schists, the Permian Gondwana sandstone for motions of Denwa, Bagra, Deccan trap sills and Dykes and also recent Alluvium characterizes the rock and soil in the region. It has typical monsoon climate with three distinct seasons. Summer, rains and winter. Variation in climate also occurs from place to place. Pachmarhi plateau is cool in summer and also has heavy rainfall during rainy season. Winter is quite cool. Lowlands in Narmada basin are uncomfortably hot in summer with less rainfall and warmer winter.

Based on land sat imagery of 1990-92 the land use of the area comprises forest (65.2 per cent) agriculture (27.7 percent) water bodies (4.2 per cent) wasteland (2.5 per cent) and built up land (0.5 per cent). Out of the total forest cover the closed forest constitutes 85.3 per cent, open forest 8.2 per cent, degraded forest 4.2 per cent and forest blank 2.3 per cent. Of the total area covered under water bodies 30.2 percent fall under river/stream and the rest 69.82 percent falls under reservoir and tanks. Tawa reservoir is the most important water body in the area. Important rivers are Dudhi, Tawa and Denwa all of which flow north and drain into river Narmada.

The Bio- Reserve is rich in plant diversity and *gene* pool. About 30 species of Tallophytes, 83 species of Bryophytes belonging to 16 families, 7 species of Gymnosperms belonging to 3 families, 1190 species of Angiosperms of 127 families and 633 genera have been reported in the area. It is a natural junction of two most important timber species, teak and Sal. The western end of Sal forest meets the teak forest in this reserve. It is rightly commented as Botanists paradise. The entire forest can be classified into three major types. However, with changing micro climatic conditions and topographical features association of plant species dominating tree species, the forest can be classified into seven types. The most important feature is the occurrence of Sal forest in the predominantly teak region makes the area unique.

Pachmarhi Bio-Reserve and the area around is largely an agricultural rural setting with preponderance of forest in the BR. It is indicative of winter land situation. The tribals overwhelmingly dominate the population in the core as well as buffer zone. Previously they used to practice shifting Dahya cultivation following slash and burn method. It has been banned in early part of the last century and they now rely on settled plough cultivation. Besides farming, they rear cattle, goat and poultry birds. Working in the forest department and collecting minor forest products are the supplementary source of income for the villages in the project areas. Their economy revolves around the forest. Even for the cure of the sick as well as prevention of seasonal illness they rely more on the Parihar or Bhopa or the medicine man-cum-shaman and wild plants and trees. Medicinal use of herbs is quite common in the interior and far flung villages. Modern medical aid is scarce due to poor transport facilities available to the local inhabitants.

It was in the latter half of twentieth century Pachmarhi has evolved as a destination for the nature tourist though it was discovered in the nineteenth century under the British rule. But it was in the map of sacred centre for the Hindu pilgrims for centuries. Several places have been associated with the great epics. Pandav caves are the reminiscence of the residence of the mighty Pandavas. Presence of rock paintings a few of which date back to 1000 A.D and even earlier have given archaeological significance to Pachmarhi.

Gond, group of tribes, Korku/Mwabi and Bharti or Bharia are the tribal groups living in the core and buffer zones. Around fourteenth century the area has come to be called *Gondwana* after the name of Gond who were the dominant component of the tribal resident. During fourteenth and fifteenth century many Rajput groups made inroads into this area and probably had succeeded in reclaiming the Narmada valley for agriculture. The tribes retired to the higher plateau and slopes to continue their hunting and own method of *Dehiya* or shifting cultivation. The tribals have maintained their ethnic boundaries and project diverse ethnic tradition.

Pachmarhi plateau is a Botanist's paradise due to its rich plant diversity. Presence of deep gorges on the Pachmarhi plateau have resulted in creation of several water falls, marshy places, perennial streams and hills of various heights. Such streams and dark shady gorges which receive sunrays for very limited hours in a day have resulted in growth of several moisture loving species of ferns, orchids, algae and many tiny herbs of immense ecological and economical importance. There are some localized important species may be considered as region specific. The gene pool rich areas are visited by tourists and students regularly as well as by local medicinal plant collectors.

The great diversity of eco-morphology and vegetation give rise to several habitats and ecological set up which support rich wildlife. All the three reserves have a much better population of wildlife as compared to other reserved forest areas. Forsyth has reported the presence of lion, elephant, wild buffaloes, etc. most of which have disappeared now. Even then over fifty species of mammals, large variety of birds, thirty species of reptiles and a host of other forms of animals are found in this BR. These forests have both grey as well as the red jungle fowl which are usually found separately in north or south India. Tigers, panthers, wild boar, barking deer, cheetah, Sambhars, etc, are the important wild life of Pachmarhi. Submergence of Tawa reservoir and Denwa river provided ample habitat for fishes, crocodile and other acquatic elements.

The landscape of Pachmarhi plateau is a tourists' delight and a hill station par excellence for the people looking for a peaceful retreat from the busy urban schedule. The hill tops and slopes have a mantle of vegetation while the plains on the plateau are vast open grassy meadows.

There are a couple of cave shelters of great archaeological interests having a number of rock paintings executed by the tribal people. Majority of these paintings belong to historical age being 2500 to 1500 years old. Mahadeo Catacomb, Jatashankar, Pandav caves and Mandia deo are important because of rock paintings from Archaeologists' viewpoint.The paintings depict hunters, animals, weapons, etc.

All these multifaceted characters have contributed towards evolving Pachmarhi as an impeccable destination for the tourists interested in nature and also an ideal place for eco-tourism. Since mid twentieth century Pachmarhi has become an important spot in the tourist map of central India. During last few decades' tourism as a whole in the country and particularly at Pachmarhi has taken the shape of an industry and a source of resource flow.

Even prior to this boon in tourism Pachmarhi was destination for the pilgrims or religious tourists. The land was bestowed by the visit of the Pandavas of the great epic. Mahadeo cave and its propitiating deity lord Mahadeo or Shiva, Mahadeo of Chauragarh the second highest peak in Satpura range have made it a sacred place and a centre for pilgrimage. Associated with this is Handi Kho, a deep gorge is believed to be retreat of a monstrous serpent, Jambudweep is another gorge which has a cavern in the rock cut by water action to resemble a giant lock of matted hair famous as Jata shanker or matted hair of lord Shiva. In geological terms these are stalactites and stalagmites. This was undoubtedly a place of Shiva worship which attracted pilgrims from the tribals as well as non-tribals. The hills of Mahadeo are made of soft rock which is cut into deep gullies by water evasion creating water falls indiscriminately distributed all around Pachmarhi plateau. There are rifts like Patalkot near Tamia in Chhindwara district having twelve hamlets or small villages of Bharia tribesmen which is approachable by a footpath only and to reach there one has to descent very carefully along the steep footpath by around mid day. Sunrays reach the floor around 10 a.m. and by three p.m. the shadow extends from the horizon.

Chauragad Mahadev and Bara Mahadev in caves are considered as most sacred. In a calendar year there remains fair and pilgrim flow on four occasions. Rakshabandhan, Deouthniegaras, Nagpanchmi and Mahashivaratri. The latter two are most important and Mahashivratri attracts maximum number of religious tourists. On Nagpanchmi day the pilgrims follow the traditional route via Almod to reach Nagdwari and complete twelve kilometres Parikrama (circumscribe) on a fixed route touching several temples and places. In Mahashivratri they visit three main temples Chaurababa, Gupt Mahadev and Bara Mahadev. Road network has developed and now the pilgrim can reach upto Bara Mahadev by road. Jeep, cars on hire are readily available. There is a stair case to reach Chauragarh from Bara Mahadev. The traditional route via Nandya is not only very tough but extremely risky. One has to walk several kilometres in this route due to poor condition of road and non-availability of vehicles. Besides one has to scale a height of more than 1000 metres along a pathway to reach these temples. Mahadev on the other hand is ten kilometres from Pachmahri connected by all weather roads. Jeep, taxi and mini bus ply on a regular basis catering to the needs of the visitors round the year. From Mahadeo cave a little ahead, a staircase is there leading to Chaura Garh temple. In spite of these available facilities the pilgrims follow the traditional route during Mahashivaratri festival. This way they satisfy the concept of salvation through life negation or through sufferings and hardship deeply embedded in the tradition and psyche. This drives them to follow the old route for pilgrimage.

There is no denial of fact that pilgrimage may be considered as religious tourism which requires certain orientation by the local tribals to manage the mega affair. Such a large influx of pilgrims tribals and caste people including urbanites demand management at the grass route level. This is however operative for centuries along this route. It is to be noted here that both Chauragarh and Mahadev cave were under the control of local Korku rulers. Sacred performers were of a different kinds than the pan Indian sacred centre traditions. Brahmanical dominance was conspicuous by its absence. Of late one Akhada or saint organization of Shaiva order is gradually stepping in. Already they have taken the control of Mahadev cave so far as the management and maintenance is concerned. A saint seat has been installed who supervise the various activities. It seems that there has been an addition of a different kind of tourism at Pachmarhi like nature tourism or eco-tourism to its established pilgrim destination. The present endeavor is to understand the traditional pattern of orientation which has evolved through and existed for centuries at Pachmarhi. It can be said with a caution that most of our pilgrim centres are also the places of nature's delight.

At this juncture it is to be said that for the present study we have selected a cluster of seven villages which come under Satpure National Park and Pachmarchi Sanctuary. The village Binora comes under Satpura National Park which falls under Core Zone while the rest six villages belong to Pachmarhi Sanctuary. These villages are adjacent one of which i.e. Nandya is most important being the last camp for the pilgrims. They start climbing the hill from this village. Nandya is also the Panchayat headquarter for the group Gram Panchayat for this cluster of villages. This cluster of villages comes under Hoshangabad district bounded by river Denwa which rises from Dhupgarh, the highest peak in the Satpura range. Villages on the south and east of this river administratively belong to Chhindwara district. The districts of Hoshangabad, Chhindwara, Balaghat, Mandla, Betul constituting Mahakoshal area dominated by the tribals. The villages under study reflect the same trend and have total tribal population. Korku/Mwasi Thakur, Gond groups and Bharia/Bharti are the tribals populating these villages. This cluster was under a Karku or Mwasi Thaskur state of Jhont.

The ruins of old royal palace, stands as the historical relics. The descendants still live in that village. The serpentine course of river Denwa separates villages of two districts. It has a perennial flow which people usually cross on foot round the year except during rains. But this flow remains intermittently allowing space to cross. During rains the river remains in full spate. The Korku and Gond are the two major tribal groups in the region. Korku occupy the western and southern parts while the Gond populated villages are common on eastern and southern parts. Bharti are a minority group whose villages are near Tamia and Patalkot.

Nandya is located at the foot of the hill range on top of which are the two important temples, Mahadeo and Chauragarh. Chauragarh is visible by the naked eye. All the three yatras during Shivaratri, Rakshabandhan and Dev Uthni Egaraes have special significance to Nandya as it is on the yatra route. On the occasion of Nagpanchami yatra the pilgrims take the route via Almod. Rorighat to reach Nagdwari and take Parikrama(circumvent).

In the light of these facts it was decided to conduct the present study in this cluster of villages centering Nandya. Mahadeva cave temple used to be managed by the other Korku state of Dabka near Matkuli. Hence the Korku are more attached to this temple trio Chaurababa Bara Mahadev (cave) and Gupt Mahadev which are visited on the eve of Mahashivratri. On Rakshabandhan and Dev Uthni Eagaraes the pilgrims visit Bara Mahadeo only. Nandya and adjoining villages have Korku as the major population. A group of Bharti is there at Nandya who have migrated to this village for long back.

Pachmarhi is the sub-divisional headquarter. Buses are available from Hoshangabad and state capital Bhopal at regular intervals. Else one can reach Piparia by train which is on the trunk route from Bhopal, Mumbai or Kolkata and from Piparia buses are available. The shortest route to Nandya is from Pachmarhi to Mahadev cave by minibus and from there one has to descend along the steep pathway to Nandya. This pathway is approximately six kilometer long. The road route is Pachmarhi to Matkuli (30 kms), Matkuli to Kuabadla via Delakhari (40 kms) and Kuabadla to Sangakheda (22 kms). One has to walk six kilometres crossing river Denwa along the fair weather forest road to reach Nandya.

Sangakheda to Nandya one has to walk or hire a bullock cart (depending on availability). Matkuhi to Kuabadla local buses are available though frequency is very poor. Occasionally jeeps are available from Delakhari to Sangakheda. Alternative route is Malkuli to Sitadungri (26 kms) and from Sitadungri fair weather forest pathway is there. One has to cross Denwa river before reaching village Churni. The distance from Sitadungri to Nandya is 29 Kilometre. Five of the seven villages of this cluster are located on this route. After Sitadungri Churni comes and then Sankhri. Both of these are Gond villages. After Sankhri other villages come one after another. First Sukhdongar followed by Digri and Gutkheda. All the three villages are Mwasi/Korku villages, Nandya is a multi-ethnic village from Nandya at a distance of five kilometres on the south-west direction, village Binora is located which is Korku village. Except Nandya all the six villages are uni-ethnic in composition. The pilgrim comes mainly from Mahakoshal area and from Vidarbha part of Maharashtra. They come to Chhindwara to Tamia. From these they come to Kuabadla by bus. From Kuabadla they reach Kheda on foot and proceed to Nandya. The alternate route is via Damua, Jhont to Sangakheda and to Nandya. Tamia and Damua are first camp sites. All come to Sangakheda the holy place of Bhura Bhagat. They seek blessings from Bhura Bhagat which is a must for every pilgrim. Then they may camp at Nandya which is last camp site and main venue for the Shivratri fair. From Nandya commences the last leg of yatra i.e. climbing the hill to reach the temple of Choura Baba. So Nandya is most important for this yatra.

Nandya is a small village located in a pit surrounded by ranges of different elevation. On its north is the Chauragarh range standing like a wall touching the sky. The settlement is a combination of linear cum dispersed hutments. There is one linear row stretching from east to west and then dispersed huts. Each hut comprises of one all purpose hut with corridor on either or both sides. The herth is inside the hut which is bisected by placing grain storage bins along the centre. Walls are made of processed mud with wooden poles supporting the roof. Roofs are double sloped and country tiled. Every hut has an entry door and a door going out. The entry door is usually on the north.

Nandya is the headquarter for group Gram Panchayat. The Panchayat office is on top of a rock at the entrance of the village. This is also headquarter for Nandya and Binora circle of the Forest Department. Forest check post is there as well as the residence of the forest guard for both the circles. Nandya forester is stationed here having two other employees for the maintenance of the wireless and check post. Nandya circle comes under Churni looked after by a Deputy Ranger. The Nandya circle in-charge is a Deputy Ranger waiting for posting. This check post is near the Nandyadev at the entry of the village. The village primary school has two teachers. Energy connection is there though the power supply is very irregular. Nagpur trustee with it's headquarter at Itwari area of Nagpur has a Dharmshala (free accommodation for pilgrims), pilgrim Rest house). Accommodation is provided to the member pilgrims. Besides it holds Bhandara or offers food free of cost to the pilgrims. Another Dharmashala is under construction.

There are two hand pumps in Nandya. People depend more on the stream flowing along the southern boundary of the village. It has a perennial flow and clean water. At the upper stretch people use this water for drinking. It meets river Denwa about half a kilometre ahead. This stream is important for the villager as they take bath in the flowing water. All the indigenous distilleries are set along the course of this stream. However, for irrigation purpose they hardly use this source. A small anikut is there and a reservoir which is kept at full capacity during Shivratri fair.

The site of the fair stretches from Nandya deo to the base of the hill along the pathway. Shops and temporary resting places are erected on either side of the pathway. This is the main settlement having linear row of huts. Fair price shops have no building. The dealer brings the items and store at the forest check post and distribute within two to five days. Medical facilities are available at Delakhari. One sister (midwife) is there for this panchayat who visits on regular intervals or on demand.

Nandya is located in a depression and the land is wavy. Cultivation is done mostly on the slopes with some occasional patches of plain plots. It is a Bharti village with a sizeable Mwasi/ Korku tribal population. This is the only village in the vicinity having Bharti or Bharia tribesmen. The Kotwal of the Panchayat is a Bharti from this village. Gopilal and Bihari both are Bharia who serve as Parihar or medicine man. They also make offerings to the village deities. Dasarath and her Bharia from this village serve as Bhumka or Bhagat or the priest serving other villages of this panch also. The Kotwal Atari who is a Bharia presides over the traditional panch.

Nandya has a population of 207 distributed into Korku and Bhania tribes. Out of a total of 3 households 21 belong to Bharia or Bharti and 17 to Mwasi/Koku. Forty five male and forty one females comprise the total Korku population of 86 souls. Among the Bharti out of total of 121 there are 58 males and 63 females.

Table 3.1
Tribe-wise Population Distribution by Age and Sex (Nandya)

	Korku		*Bharti*		*Total*	
Age break-up	*Male*	*Female*	*Male*	*Female*	*Male*	*Female*
0 - 5	13	10	15	14	28	24
06 -- 10	7	4	7	11	14	15
11 -- 15	4	9	3	5	7	14
16 -- 20	3	0	7	4	10	4
21 -- 25	1	3	4	8	5	11
26 -- 30	5	3	9	7	14	10
31 - 35	4	7	4	2	8	9
36 - 40	4	1	2	2	6	3
41 - 45	1	1	2	2	3	3
46 - 50	0	0	1	7	1	7
51 - 55	2	1	3	0	5	1
56 +	1	2	1	1	2	3
	45	**41**	**58**	**63**	**103**	**104**
Total	**86**		**121**		**207**	

It is to be stated here that the Bharia or Bharti are considered as low by the Mwasi Thakur. The Bharti are believed to consume buffalo meat which has put them a status lower than that of Gond. Mwasi enjoy the highest status being the ruling group. They maintain commensal distance with the Bharti though they accept their services. It seems to be original Bharti village where Mwasi

Table 3.2

Literacy Rates (Nandya)

Tribe	*Age breakup in years*	*Illiterate*					*Literacy by standard*				
		Male	*Female*	*Male*	*Female*	*Male*	*Female*	*Male*	*Female*	*Male*	*Female*
Bharti/	00 - 05	12	7	1	1	-	-	-	-	-	-
Bharia	0 6- 12	-	2	7	8	1	-	-	-	-	-
	13-18	2	2	1	3	3	-	-	-	-	-
	19 - 25	-	3	1	-	-	-	-	-	-	-
	26 - 30	3	3	2	-	-	-	-	-	-	-
	31 +	9	11	2	1	1	-	-	-	-	-
	Total	**26**	**28**	**14**	**13**	**5**	**-**	**-**	**-**	**-**	**-**

are immigrants. In the old records Nandya, Thela and Kanjighat have been shown as Bharti villages. However the Mwasi Thakur are the dominant group in the area in terms of population strength. Also there are Gond villages. Population of the female is little higher than the male among the Bharti. Both Bharti and Mwasi Thakur speak in local dialect of Hindi and have lost their ethnic dialect. The Mwasi even have no idea of their tribal dialect.

The literacy attainment among the Mwasi is very poor exhibited by the fact that 54 out of 86 persons have been reported as illiterate. Four persons have reached secondary level of education and a lone person of older age group had achieved this level. There were none among the females who have secondary level education. Among the Bharti the picture is further gloomy, 91 persons out of a total population of 121 have returned as illiterates. Majority of them are females. None have attained secondary level of education. At the primary level there are only twelve females out of 30 individuals. The Table 3.2 depicts that in both the tribal groups' literacy level is very poor. Further the females are lagging far behind among the few who have taken some education. It is also evident that older generations are mostly illiterate. The silver line is in both the cases they have started sending their children to school. Primary school at the village have two post of teachers. One has recently been transferred to Sukhdongar. But he lives here and commutes daily. The teachers posted in the nearby villages very often visit the village. One of the teachers is a Gond from Chutri village and another is a Mwasi who takes keen interest in the students. The forest officials cooperate with the teachers posted here. The Nakedar or incharge of the checkpost is a Gond. The Nakedar has two assistants one is a Gond from a village near Tamia and the other is a Mwasi/ Korku from this village. The quarters are pucca houses and as the forester is living alone the teachers after meal in the evening come here gossip and sleep. The fare price shop dealers also camp here. So is the case of Panchayat secretary when he requires to stay back.

The villagers primarily depend on agricultural farming. Lands are mainly on slopes with occasional plain and leveled patches. Irrigation facilities are not there and resources have not been

Table 3.3
Occupational Distribution (Nandya)

Age break up in years	Farming		Daily Wage		Service		Forest Collection		Distillation		Misc.	
	Male	Female	Male	Female	Male	Female	Male	Female	Male	Female	Male	Female
15 - 30	2	2	5	4	1	-	-	-	1	1	-	-
31 - 40	2	3	2	3	1	2	-	-	3	-	-	-
41 - 50	1	1	-	-	-	-	-	-	-	-	-	-
51 - 60	1	1	-	-	-	-	-	-	1	-	-	-
61 +	1	1	-	-	-	-	-	-	-	-	-	-
Total	**7**	**8**	**7**	**7**	**2**	**2**	**-**	**-**	**5**	**1**	**-**	**-**

Table 3.4
Occupational Distribution (Nandya)
Tribe - Bharti

Age break up in years	Farming		Daily Wage		Service		Forest Collection		Distillation		Misc.	
	Male	Female	Male	Female	Male	Female	Male	Female	Male	Female	Male	Female
15 - 30	2	2	5	4	1	-	-	-	1	1	-	-
31 - 40	2	3	2	3	1	2	-	-	3	-	-	-
41 - 50	1	1	-	-	-	-	-	-	-	-	-	-
51 - 60	1	1	-	-	-	-	-	-	1	-	-	-
61 +	1	1	-	-	-	-	-	-	-	-	-	-
Total	**7**	**8**	**7**	**7**	**2**	**2**	**-**	**-**	**5**	**1**	**-**	**-**

tapped. Cultivation is purely dependent on climatic condition particularly monsoon rains. Usually single crop land where maize, pulses and oilseeds are grown. Besides they grow wild variety of cereals like Kodo and Kutki. These were originally grown by the tribals in this area along with Rajgira. Kodo they prepare like rice and consume with pulses. Maize cultivation is a recent introduction. The combined output from the agricultural operative is far from sufficient for the living of the resident. They depend on the forest to make up the deficit. Forest provide them lot of resources. They get the fire wood free of cost and sell the surplus during fair. Also they collect minor products from the forest like Lahenja, Kossam fruits, honey, Menhar, Bil, *achar* etc. Mohul leaf plucked to prepare leaf plates and cups. Prior to the imposition of ban they used to sew leaf plates of *sal* leaves. The bark of Aarjan is used to make rope. Tendu leaf used for biri rolling is collected on wage basis. Papda, Dhokda and Kani are collected for fire wood for self consumption and also to sell during Shivratri fair. Also they collect a host of wild fruits. Mahua is an important commodity they collect for own consumption and sell too. There are several preparations of Mahua flower and also used for distilling liquor. Barks, fruits, roots, leaves etc, of a number of plants are collected for medicinal purposes. Indigenous use is there and also they collect for Ayurvedic Medicine manufactures.

In every house there are goats and poultry birds. This fetches good return in the village as well as in every Tuesday's weekly market at Sangakheda. Besides, during yatra goat and poultry bird are the most sought after commodity for sacrifice at Chaura baba. The sacrifice is essential. It may not be necessarily killed rather left in the name of the deity. The cost of these in those days goes up. The other important commodity is Mahua liquor for the pilgrims.

This apart they go for daily wage in forest department or road construction. During the harvest of winter crop there remains a demand for additional labour in the plain areas near Shobagpur, Piparya area. The area in question has been changed due to Tawa Hydel project. Irrigation facilities have brought about a boon in farming and created the requirement of a large quantity of man hours. The tribals who can manage go there on contract basis to work in harvest operations go for a higher rate of wage.

Under the scheme of Joint Forest Management like the neighbouring villages of this cluster there is an Eco-development society at Nandya. This village covers under buffer zone. It's a revenue village having 890.30 hectares of forest land. The society is responsible for protection from forest fire, from illegal cutting of live trees, illegal transportation, illegal grazing, illegal mining, encroachment and desist hunting of wild animals. They are to inform and assist the forest department. Protection of estranged wild life and to inform forest department, informing the nearest forest office about illegal intruders in the forest and finally protection of forest products are part of duty of the tribal inhabitants. An amount of four to five thousand rupees are granted to the members at 2 per cent per month rate of interest. This rate is much cheaper than what is charged by the local moneylender.

Gutkheda is located on the east of Nandya at an approximate distance of four kilometres. This is totally a Mwasi/Korku village. The main settlements have linear rows of huts on either side of the main road passing through the settlement. There is a by pass too. Other huts are dispersed being farm huts. Hand pumps are there for drinking water. The lower primary school building is in ruins. So the classes are being held in the verandah of Suraj Patel. The single teacher school is regular in holding classes. Suraj has provided accommodation to the teacher free of cost and also free food. He is the Patel of the village. Nawal Singh is the Bhagat or the priest. Soratiabai of Mwasi/Korku tribe is the Sarpanch of the Group Gram Panchayat Nandya. The seat is reserved for a lady candidate.

In Gutkheda there are 24 Korku households having a total population of 150 people of which 80 are males and 70 are females. Nuclear type of family is most common followed by vertically extended family types. There is a forest check post. Northern boundary of the village is marked by the forest fair weather road. Pachmarhi ranges stands as a wall on the northern horizon. Agriculture is the main stay of the village. Farm lands are in small patches and scanty. Bullock cart is the usual vehicle for transport.

Literacy level is very low. Only less than one third of the population is literate. None have reached secondary level of education. It has become evident that the little progress which

Table 3.5
Tribe-wise Population Distribution by Age and Sex
(Gutkheda)

	Korku		*Total*	
Age break-up	*Male*	*Female*	*Male*	*Female*
0 -- 5	21	12	21	12
06 -- 10	12	8	12	8
11 -- 15	11	7	11	7
16 -- 20	4	9	4	9
21 -- 25	7	5	7	5
26 -- 30	9	9	9	9
31 - 35	6	6	6	6
36 - 40	2	2	2	2
41 - 45	2	3	2	3
46 - 50	4	2	4	2
51 - 55	1	3	1	3
56 +	1	4	1	4
	80	**70**	**80**	**70**
Total	**150**		**150**	

has been made is a recent development. Older people have totally been returned as illiterates. Though the village has given the first lady sarpanch to the group Gram panchayat, Nandya the female literacy performance is very poor.

Occupational pattern shows that they are primarily dependent on farming and daily wage. None in the village is in government jobs. They collect firewood and minor products from forest. The village is uniethnic with total Korku/ Mwasi people. They are attached to the *yatras* being adjacent to village of Nandya. They make a ready stock of Mahua liquor, fire wood, chicken and goat to avail the pilgrim customer. Sixteen of them opened shop at Nandya during Shivratri yatra. In group they took the car parking and barrier at Nandya on contract. Preceding year a group of five persons from Matkuli took this contract and fetched good profit. This year they put their claim strongly on the plea that they are localites. The panchayat upheld their claim along with those of other neighbouring villages.

Table 3.6
Literacy Rates (Gutkheda)

	Age break up in years	*Farming*		*Daily Wage*		*Service*		*Forest Collection*		*Misc.*	
		Male	*Female*	*Male*	*Female*	*Male*	*Female*	*Male*	*Female*	*Male*	*Female*
Korku/	00 - 05	21	12	-	-	-	-	-	-	-	-
Mawasi	0 6- 12	-	4	14	9	-	-	-	-	-	-
Thakur	13-18	-	2	11	4	-	-	-	-	-	-
	19 - 25	6	7	2	3	-	-	-	-	-	-
	26 - 30	6	8	3	1	-	-	-	-	-	-
	31 +	18	19	0	-	-	-	-	-	-	-
	Total	**51**	**52**	**30**	**17**	-	-	-	-	-	-

Table 3.7
Occupational Distribution (Gutkheda)
Tribe-Korku

Age breakup in years	*Farming*		*Daily Wage*		*Service*		*Forest Collection*		*Distillation*		*Misc.*	
	Male	*Female*	*Male*	*Female*	*Male*	*Female*	*Male*	*Female*	*Male*	*Female*	*Male*	*Female*
15 - 30	7	8	11	15	-	-	-	-	-	-	-	-
31 - 40	4	6	3	2	-	-	-	-	-	-	-	-
41 - 50	5	4	1	2	-	-	-	-	-	-	-	-
51 - 60	1	2	-	1	-	-	-	-	-	-	-	-
61 +	1	3	-	1	-	-	-	-	-	-	-	-
Total	**18**	**23**	**15**	**21**	**0**	**-**	**-**	**-**	**-**	**-**	**-**	**-**

The villagers have darshan of Chaurababa, Mahadev and Gupta Mahadev about one week earlier so that their programme does not clash with the schedule of the fair. Though the Shivratri day is most auspicious to have darshan of lord Shiva, they had to compromise as they get included in the fair in a big way. It's a regular practice for the Mwasi of this cluster of villages.

Gutkheda is devoid of any public buildings. Village Patel presides over the village meetings. The post is hereditary like that of village Bhagat or the priest. Bhagat makes the offerings to the village deities from time to time on fixed occasions. The village Parihar is the medicine man.

Digri is another Mawasi village on the south east direction of Gutkheda and Nandya spreading along the course of river Denwa. Denwa forms the southern boundary of the village. Land is fertile and farmers use perennial flow of water of Denwa for irrigation. Agriculture is the main source of livelihood. In addition to rain crops they cultivate winter crops like wheat. The village primary school runs in the house of the Patel. There is no school building. It is a single teacher school and the teacher lives at Nandya from where he comes to this village daily on forest. The villagers are ready to provide him accommodation. But he not accept as the village is a wet village. After dusk it is difficult to stay. The villagers are not enthusiastic to have a good school building. Government buildings are conspicuous by their absence in the village. Here too linear arrangement of hut is there alongwith a few isolated, dispersed huts. The walls are made of mud and the frame is erected by wooden poles and bamboo. Double sloped country tiled roof is the set pattern. The village is on the slope of a hillock and river Denwa flows in a semi-circular fashion. On the other side of the river is the village Sangakheda which is also a Mwasi village. Sangakheda is a large village with dispersed settlement. It is an important village being the camp site for pilgrims. Bhurabhagat temple has added to its sacredness. The weekly market is held here on every Tuesday. Proximity of Digri to Sanghkheda has contributed positively. The later is well connected with Kuabadla, Matkuli, Piparya and district towns Chhindwara and Hoshangabad by all weather roads.

Digri has its own Patel and Bhagat. The village is represented in the Nandya Panchayat by a member. They visit

Nandya as well as Sangakheda frequently for day to day necessities.The yatra are important for the villagers. It is a great occasion of the year. The village has 19 households belonging to Mwasi/Korku tribe. The total population is 114 having 62 males and 52 females. There is only one female in 50+ age group.

Only twenty seven persons are literate in the village of which five are females. Large section of the females are illiterate and five female literates have attained primary level of education only. Three boys are studying in Ashram school at Pachmarhi. The school remains open irregularly facilitating the children to enjoy playing or working in family farm. Only one male from 31 plus age group has returned as having primary level of education.

Village economy revolves around farming. Kharif crops like maize, kodo, *kutki* are grown and the cycle of operation is dependent on the rainfall. River Denwa has a perennial flow and the farmers here use this resource for Ravi crops. River side farm lands are fertile and give good yield. Obviously a large chunk of the working force is involved in farming as primary source of livelihood.

The villagers are also economically tied up with the Shivratri yatra. So they make a stock of Mahua liquor long ahead of the Shivratri day. Rearing goat and poultry bird is common among them. It is part of their life. On several occasions in life cycle rituals they sacrifice chicken and goat. So in every house there will be a goat and poultry bird. The they adopt is that they have a routine depending on the growth rate (life cycle) and food intake they raise the chicken. This they do in such a way that they will be full grown and ready for sacrifice around Shivratri fair. The scheduling actually starts around Dev Uthni eqaras (October) for poultry bird and one year ahead for a goat. For fire wood they start collecting from forest immediately after the rains. They then split from time to time and leave it to dry further in the sun. Distillation of Mahua liquor they start two months earlier so as to have a ready stock of five hundred litres or more by the time yatra begins. A little delay cause foregoing their trip to plain areas during the harvest of Ravi crop. This harvest trip is of reasonably less significance than Shivratri fair. This provides employment as guide, porter, physical assistants for scaling the height by an aged or weak pilgrim. Besides, this year twenty of them have opened shop along the route of the pilgrimage.

Table 3.8
Literacy Rates (Digri)

Tribe	*Age break up in years*	*Illiterate*		*Literacy by standard*							
		Male	*Female*	*Male*	*Female*	*Male*	*Female*	*Male*	*Female*	*Male*	*Female*
Korku/	00 - 05	14	7	1	-	-	-	-	-	-	-
Mawasi	06- 12	3	6	12	2	-	-	-	-	-	-
Thakur	13-18	3	7	2	3	3	-	-	-	-	-
	19 - 25	7	11	3	-	-	-	-	-	-	-
	26 - 30	4	1	-	-	-	-	-	-	-	-
	31 +	11	15	1	-	-	-	-	-	-	-
	Total	**42**	**47**	**19**	**5**	**3**	**-**	**-**	**-**	**-**	

Table 3.9

Occupational Distribution (Digri)

Tribe - Korku

Age break up in years	*Farming*		*Daily Wage*		*Service*		*Forest Collection*		*Distillation*		*Misc.*	
	Male	*Female*	*Male*	*Female*	*Male*	*Female*	*Male*	*Female*	*Male*	*Female*	*Male*	*Female*
15 - 30	14	10	5	5	-	-	-	-	2	2	-	-
31 - 40	5	6	-	-	-	-	-	-	-	1	-	-
41 - 50	7	7	-	-	-	-	-	-	-	-	-	-
51 - 60	-	-	-	-	-	-	-	-	-	-	-	-
61 +	-	1	-	-	-	-	-	-	-	-	-	-
Total	**26**	**24**	**5**	**5**	-	-	-	-	**2**	**3**	-	-

Sukhdongar is a little large village four kilometre east of Gutkheda. It sprawls over a kilometre and has three hamlets at different heights. Linear rows of huts are arranged in east west direction. Both from internal arrangement and external look these huts are almost replica of those of Digri and Gutkheda. The village lanes are along the elevation. Village primary school building is in the hamlet located at the lowest lever. The fair weather forest road passes through this hamlet connecting this village with neighbouring ones. Pachmarhi range stands as a wall forming the northern boundary. Top of it projects the shape of elephant tusk so referred as Hatimatha range. The village is located at the foot of this range. Cultivable lands are mainly in the slopes. They grow kharif crops which is dependent on normal cycle of monsoon rains. Each house has goats and poultry birds. It is a uniethnic village having fifty four Mwasi/Korku households.

Table 3.10
Tribe-wise Population Distribution by Age and Sex (Sukhdongar)

	Korku		*Bharti*		*Total*	
Age break-up	*Male*	*Female*	*Male*	*Female*	*Male*	*Female*
0 - 5	40	33	0	1	40	34
06 - 10	23	22	0	0	23	22
11 - 15	15	14	1	0	16	14
16 - 20	7	15	0	1	8	15
21 - 25	17	21	0	0	17	21
26 - 30	17	9	0	1	17	10
31 - 35	9	8	1	0	10	8
36 - 40	11	8	0	0	11	8
41 - 45	4	5	0	1	4	6
46 - 50	4	5	0	0	4	5
51 - 55	4	4	1	0	5	4
56 +	1	6	0	0	1	6
	152	150	3	4	156	153
Total	**302**		**7**		**309**	

The total population is 309 distributed to 155 males and 154 females. Population wise it is a large village as compared to other

Table 3.11
Literacy Rates (Sukhdongar)

Tribe	*Age break up in years*	*Illiterate*		*Literacy by standard*							
		Male	*Female*	*Male*	*Female*	*Male*	*Female*	*Male*	*Female*	*Male*	*Female*
Korku/	00 - 05	40	34	-	0	-	-	-	-	-	-
Mawasi	0 6- 12	11	3	19	25	-	-	-	-	-	-
	13-18	8	7	2	2	-	1	-	-	-	-
	19 - 25	21	34	-	-	-	-	-	-	-	-
	26 - 30	15	9	3	-	-	-	-	-	-	-
	31 +	35	36	0	-	-	-	-	-	-	-
	Total	**130**	**123**	**24**	**27**		**1**	**-**	**-**	**-**	

Table 3.12
Occupational Distribution (Sukhdongar)
Tribe – Korku

Age break	*Farming*		*Daily Wage*		*Service*		*Forest Collection*		*Distillation*		*Misc.*	
up in years	*Male*	*Female*	*Male*	*Female*	*Male*	*Female*	*Male*	*Female*	*Male*	*Female*	*Male*	*Female*
15 - 30	17	18	20	21		1	-	-	4	4	-	3
31 - 40	14	9	7	6	-	1	-	-	-	-	-	-
41 - 50	4	7	2	5	-	-	-	-	2	-	-	-
51 - 60	4	3	1	-	-	-	-	-	-	-	-	-
61 +	1	3	-	1	-	-	-	-	-	-	-	-
Total	**40**	**40**	**30**	**33**		**2**	**-**	**-**	**6**	**4**	**0**	**3**

neighbouring villages. Nuclear type of family is common among them. The village primary school has two teachers one is a Korku and another is a lady teacher belonging to Gadi caste. Like that of Nandya their school provides mid-day meals to the students. The literacy rate is still very low. Only seventeen percent of the population is literate and only one girl has reached secondary level of education.

Table 3.13
Tribe-wise Population Distribution by Age and Sex (Sankhri)

	Gond	
Age break-up	*Male*	*Female*
0 - 5	11	12
06 - 10	5	9
11 - 15	6	2
16 - 20	2	6
21 - 25	4	4
26 - 30	4	3
31 - 35	6	5
36 - 40	2	1
41 - 45	2	0
46 - 50	1	0
51 - 55	1	2
56 +	3	3
	47	**47**
Total	**94**	

It is a farmer's village. Farming is supplemented by daily wage in forest department and otherwise. Two women are in government service. Distillation of Mahua liquor is the primary occupation of some households. Also they raise goat and poultry birds. Collecting firewood and minor forest products help them to meet their essential requirements. The villagers get directly valued in Shivratri yatra. It is around 11 kilometres from Nandya the main venue or the last camp for the pilgrims. Rakshabandhan yatra is for a day only and the condition of the fair weather roads stand as a constraint in participating closely. They being Mawasi strongly feel that they are an essential part in the Shivratri festival. It is a common belief that Lord Shiva or Bhagwan has created them. Myth of creation depicts that lord Mahadev has created Mawasi/ Korku to protect them from evil spirits.

Table 3.14
Literacy Rates (Sankri)

Tribe	*Age break*	*Illiterate*	
	up in years	*Male*	*Female*
Gond	00 - 05	11	12
	0 6- 12	5	10
	13-18	1	4
	19 - 25	3	7
	26 - 30	3	3
	31 +	14	11
	Total	**37**	**47**

On further east from Sukhdongar is the village Sankhri. It is Gond village having thirteen households.The settlement pattern shows irregular agglomeration. It is mud walled rectangular shape with double sloped country tiled roof. Each hut has a frontal courtyard. The total population of the village is ninety four distributed equally among the sexes.

The settlement pattern is of irregular conglomerate. There is no school in the village which forces the children to go to the neighbouring villages to attend school. This speaks of the poor literacy (11.46 per cent) rate in the village. In the higher age group only two males are literate. The females are totally illiterate. The economy of Sankhri revolves round agriculture. As wage labourer they work in forest department and other government agencies. They also go as seasonal labourer particularly for Rabi crop activities to the benefit areas of Tawa reservior.

Table 3.15

Occupational Distribution (Sankri)

Tribe - Gond

Age break up in years	*Farming*		*Daily Wage*		*Service*		*Forest Collection*		*Distillation*		*Misc.*	
	Male	*Female*	*Male*	*Female*	*Male*	*Female*	*Male*	*Female*	*Male*	*Female*	*Male*	*Female*
15 - 30	3	2	7	11	-	-	-	-	-	-	-	-
31 - 40	5	4	3	2	-	-	-	-	-	-	-	-
41 - 50	2	-	1	-	-	-	-	-	-	-	-	-
51 - 60	2	4	1	-	-	-	-	-	-	-	-	-
61 +	-	-	-	-	-	-	-	-	-	-	-	-
Total	12	10	12	13	-	-	-	-	-	-	-	-

They distil mahua liquor for their own consumption. Also they cannot afford to ignore the job opportunities in the agricultural sector in plains though it clashes with Shivratri fair. Firstly the village is little far from the pilgrim route and Nandya (14 kms). Besides, the Gonds are not attached to these pilgrim centres in the way the Korku are Proximity plays an important role. Further it is said that previously Gond had no entry in the Mahadev temple in the cave.

Churni is the eastern most village under Nandya Group Gram Panchayat. It is a Gond village seventeen kilometres away from its panchayat headquarter. This and the last five villages described comes under Pachmarhi Sanctuary. River Denwa marks its southern boundary where it takes a northern eastern course and separates this from Sitadungri areas. Sitadungri along the fair weather route is 15 kilometres away passing through deep forests of Sal. This area of the forest have wild games in plenty. There is a pathway leading to Delakheri. Along this pathway Delakhari is 9 kilometres. Delakhari is the nearest large urban village for this cluster of villages providing civic and transport facilities. Delakhari is located on the Pachmarhi- Chhindara Nagpur-road. The villagers regularly visit Delakheri. The weekly tribal market is held at Sangakhere and Delakheri. Sangakhere is little far from Churni than that of Delakheri. Even then the villagers visit both the weekly markets which are held on different days of the week. (Tuesday and Friday). Weekly markets are meeting points and not merely a commercial hub. Churni is a middle sized village as compared to local tribal villages in the core and border of buffer zone with 28 Gond households having 187 souls.

In the literacy level Churni has shown remarkably better performance. More than one third of (34.43%) the population has returned as literates. Nearly half of them have secondary or higher level of education. Female literacy level is also quite significant. There are two women having graduate and post graduate degree. Surprisingly there is no school in the village. This however has been compensated by its nearness to Dalakheri where there are senior secondary and also Christian Missionary Schools.

The ex-sarpanch acted as a source of inspiration for others. His family members are highly educated. The family holds the post for fifteen years since its inception and till recent election when it was declared as reserved for lady sarpanch.

Besides, Churni is an old circle headquarter of forest department. There is Deputy Rangers office under which there are two circles Binora and Nandya coming under Matkuli forest zone. The Forest Department establishment comprises of a wireless centre, colony for forest employees and a check post. On either side of this colony are the two hamlets of the village. Three hand pumps being in working conditions there provide sufficient available drinking water. Only at Churni there is a medium size temple of lord Hanuman, At other villages the monkey god is represented by one or a few vermilion smeared stones.

Table 3.16

Tribe-wise Population Distribution by Age and Sex (Churni)

	Gond	
Age break-up	*Male*	*Female*
0 - 5	14	20
06 - 10	13	15
11 - 15	11	6
16 - 20	14	7
21 - 25	9	9
26 - 30	6	6
31 - 35	5	6
36 - 40	9	8
41 - 45	4	3
46 - 50	4	4
51 - 55	0	0
56 +	7	7
	96	**91**
Total	**187**	

Agriculture is the main stay of the village economy. Here in this village the available leveled farm land is more due to more or less flat topography. Kodo, Kutki, Maize and Jawar are grown as Kharif crop. Irrigation facilities are neither harnessed nor properly available.

Table 3.17
Literacy Rates (Churni)

Tribe	Age break up in years	Illiterate				Literacy by standard					
		Male	Female	Male	Female	Male	Female	Male	Female	Male	Female
Gond	00 - 05	13	17	1	2	-	-	-	-	-	-
	06-12	4	9	14	7	1	-	-	-	-	-
	13-18	6	7	2	3	3	-	1	-	-	-
	19 - 25	7	8	4	1	3	-	2	-	-	1
	26 - 30	5	7	2	1	2	1	1	1	-	-
	31 +	16	21	5	1	1	-	1	-	1	1
	Total	**51**	**69**	**28**	**15**	**10**	**1**	**5**	**1**	**1**	**2**

The other preferred occupation is daily wage agricultural seasonal (Satam) labourers for Rabi crop. Hence the working force is almost evenly distributed between these two occupations. Seasonal labour cycle clashes with Shivaratri fair in which they give more importance to setam. Churni again a Gond village is a little aloof from the traditional pilgrim route. There are some in white collar jobs and few stick to the age old occupation of collecting minor forest products. But this is only to supplement their requirement and do not depend on this totally. Any of the three fairs are important for those who take on pilgrimage but not for economic pursuits Churni has energy connection.

Binora is the only village in this cluster falling under core zone being in Satpura National Park. It is around three and a half kilometres on the south direction from Nandya. Binora is a middle sized village totally populated by Mwasi Thakur/Korku. River Denwa flows by it encircling on three sides while on the other is Dhoopgarh range. Denwa here has a perennial flow of clear water which is successfully used for irrigation. The settlement pattern resembles Korku traditional village i.e. two linear rows of huts stretching from east to west with a common main village street in between. There are few huts constructed in agricultural farms. Farming is done in the flat slopes. The village primary school has its own building and teacher's reside little away from main settlement on an elevation overlooking river Denwa flowing down. There are *pucca* stairs in one of the potholes of the river. There are forty households having a population of 228 of which 115 are males and rest are females.

The literacy rate is quite high (38.10 per cent) as compared to other Korku villages. Interestingly the number of illiterates has been evenly distributed among the sexes. Higher education is not favoured for girls though they send their daughters to primary schools. At large very few pursue higher education.

The economy of the village is primarily agro-based. They raise common kharif crops like Kodo, Kutki, Maize and rice. Also they cultivate wheat using river water for irrigation. They collect a large number of forest products to supplement their earnings. They have a large number of wage earners getting employment in forest and public works Departments.

Table 3.18
Occupational Distribution (Churni)
Tribe – Gond

Age break up in years	*Farming*		*Daily Wage*		*Service*		*Forest Collection*		*Distillation*		*Misc.*	
	Male	*Female*	*Male*	*Female*	*Male*	*Female*	*Male*	*Female*	*Male*	*Female*	*Male*	*Female*
15 - 30	5	5	19	4	3	1	-	3	-	-	-	-
31 - 40	6	4	7	9	1	1	-	2	-	-	-	-
41 - 50	6	5	2	2	-	-	-	-	-	-	-	-
51 - 60	6	3	-	-	-	-	-	-	-	-	-	-
61 +	-	-	-	-	-	-	-	-	-	-	-	-
Total	**23**	**17**	**28**	**15**	**4**	**2**	**-**	**5**	**-**	**-**	**-**	**-**

Table 3.19
Occupational Distribution
(Binora)

	Korku	
Age break-up	*Male*	*Female*
0 - 5	32	21
06 - 10	19	21
11 - 15	10	22
16 - 20	9	10
21 - 25	11	12
26 - 30	8	5
31 - 35	6	6
36 - 40	9	9
41 - 45	6	1
46 - 50	2	2
51 - 55	0	0
56 +	3	4
	115	**113**
Total	**228**	

They are economically linked up with all the three fairs of which Shivratri is the most important. Actively they participate in Shivratri fair. It is the most important event of the year. Detail preparation is taken to built up a stock of Mahua liquor and fire wood. They raise poultry bird and goat which are normally sacrificed. As porter-cum-guide they serve the pilgrim on payment. This year thirty-one persons opened shops during the fair having good earning. Chhotu Singh of this village along with his kins and relatives took the lease of Chaurababa temple chaudatsi for eight days i.e. fair period. Lease money was 1 lakh 60 thousand which he paid in cash two years back. Priority was given to the local people and hence they were awarded the lease. Against this Garibdas, Mahant of the Akhada of Bara Mahadeo cave temple filed a suit in Bhopal High Court and obtained stay. So the district administration has stopped giving the lease and kept management under their direct supervision. For the shop near this temple they gave the bid at one lakh sixty one thousand and did not claim

Table 3.20
Literacy Rates (Binora)

Tribe	*Age break*	*Illiterate*		*Literacy by standard*							
	up in years	*Male*	*Female*	*Male*	*Female*	*Male*	*Female*	*Male*	*Female*	*Male*	*Female*
Korku/	00 - 05	30	22	-	1	-	-	-	-	-	-
Mawasi	0 6- 12	1	2	20	25	-	-	-	-	-	-
Thakur	13-18	-	4	9	20	5	-	-	-	-	-
	19 - 25	9	15	2	1	3	-	-	-	-	-
	26 - 30	8	4	-	-	-	-	-	-	-	-
	31 +	24	24	2	-	-	-	-	-	-	-
	Total	**72**	**71**	**33**	**47**	**8**	**-**	**-**	**-**	**-**	

Table 3.21
Occupational Distribution (Binora)
Tribe-Korku

Age break	*Farming*		*Daily Wage*		*Service*		*Forest Collection*		*Distillation*		*Misc.*	
up in years	*Male*	*Female*	*Male*	*Female*	*Male*	*Female*	*Male*	*Female*	*Male*	*Female*	*Male*	*Female*
15 - 30	8	7	21	24	-	-	-	-	4	3	1	1
31 - 40	6	7	8	7	-	-	-	-	1	1	-	1
41 - 50	7	2	2	1	-	-	-	-	-	-	-	-
51 - 60	-	-	-	-	-	-	-	-	-	-	-	-
61 +	1	2	1	-	-	-	-	-	-	-	-	-
Total	**22**	**18**	**32**	**32**					**5**	**4**	**1**	**2**

localite identity. It was awarded to another shopkeeper of Pachmarhi for one lakh sixty three thousand. They collect a fairly large number of forest products which they sell to the visiting traders at Nandya-or Sukhdongar. As a whole Bimova is economically better than Nandya or Digri. River water is being used by lift irrigation through pipe line to irrigate land in the upper slopes.

This village is one of those villages in the core zone marked for getting vacated. Notification in this regard has already been promulgated. Forest Department has given the following offer for rehabilitation of tribals to be evacuated from the villages like Binora in Salpura National Park. They have a plan to evacuate rest of the villages of this cluster coming under Pachmarhi Sanctuary located North of river Denwa in course of time.

The offer:

1. Cultivable plain farm land – 5 acres
2. Homestead – 0.70 acre
3. One, tiled roof constructed hut of traditional tribal pattern of this zone
4. One cattleshed attached to this hut
5. Threshing place
6. Metalled link road to main road
7. Free energy connection

The area where they will be rehabilitated is near Piparia which is plain zone outside the Bio-Reserve. Piparia is nearly ninety kilometres from Binora. The people of Binora are reluctant to vacate their village for the following facts as detailed by them. They are under pressure to leave the village and may also have to abide by the dictates of the government but would not like to leave their fore-father's village.

1. Their economy is agro-forest based supplemented by income from fairs.
2. Traditionally and technically attached to the crop pattern and work cycle in this typical terrain.
3. They collect a wide variety of minor forest products like gum, honey, Mahua flower, Tendu leaf, achar etc. Make leaf plates and cups. Collect fire wood for self consumption and sell.
4. Habituated with the food and water available

5. This is their territory where their ancestors breathed their last and also located the place of ancestral worship. They are attached to this land of their forefathers.
6. The marriage distance on an average is 15 kilometres which will not only increase several times but also require to develop new set to work.
7. Marriage alliances will be a problem.
8. Kinship network will be severely affected which is directly related to their life cycle rituals.
9. In house construction and making agricultural implements they use wood which they get free of cost from the forest. They are to forgo this privilege once out of this forest.
10. For cure of various ailments and prevention from seasonal diseases they traditionally use several herbs and roots locally available. The medical facilities available are far from the requirements they are dependent on the local pattern of treatment.
11. The villages and particularly Binora and Nandya are located at the foot of Mahadeo Hills. Hence they prefer to live here with the blessings of Lord Mahadeo believed to be their creator.
12. This cluster of village was under administrative jurisdiction of Mawasi/Korku state of Jhont. Each village has panch or council of elders and a Patel the traditional village head. This office is hereditary and follows the rule of male primogeniture regarding succession. The statutory Panchayat is also dominated by the tribals specially Mawasi being the most numerous group. In new area there will be a separate set of arbitration who have least idea about their tradition and customs.

These are some of the basic reasons which discourage the tribal villagers to shift to the rehabilitation area. Out of these several insisted that they like to remain close to their creator lord Mahadeo.

Nature tourism is a recent process in Pachmarhi as compared to age old custom of pilgrimage. Since long Pachmarhi has remained a pilgrim destination. Multiple factors contributed to the sacredness of Pachmarhi. The cave temple dedicated to lord Mahadeo is unique in the claim of its being effulgent Shivling.

Chaurababa or Mahadev in another form is there on top of second highest peak of the Satpura range. The third important temple of Mahadeo is Gupta Mahadeo. Devotees first reach Chaurababa temple then walk down the stair to Gupta Mahadeo and finish after having darshan of Bara Mahadeo in the cave. The pilgrimage is taken up on the eve of Shivratri, Rakshabandhan and Dev Uthni Egaras.

Pachmarhi is associated with the mighty Pandavas of the great epics. During their period of exile they believed to have stayed here. Jatashanker, Handikho, Nagdwari are other major shrines which have made the Botanists paradise a sacred centre too for the tribals and the Hindus. The main fair takes place on two occasions Mahashivratri and Nagpanchmi. Mahashivratri is more important. Pachmarhi Mahadeo has perennial flow of pilgrims. But Shivratri is a special occasion which draws pilgrims from far and wide area, tribals as well as caste people. Pilgrims come mainly from Maharashtra and adjacent tribal districts. Nagpanchmi is celebrated through a Parikarma (circular *yatra* touching seven main places) of sacred places in Pachmarhi plateau in a day. Pilgrims come from the plains of the south via Almod and Kajri villages. They follow the traditional pathway route. Others come by bus or other vehicle to Pachmarhi and then perform Parikarma. The pilgrims prefer traditional pathway to have darshan of the deities.

On the occasion of Shivratri fair, Rakshabandhan and Dev Uthni Egaras the traditional route or pathway is via Sengakheda and Nandya. Due to construction of road network one can reach Bara Mahadeo by vehicle. Then one has to walk a little to reach the stair leading to the temple of Chauragarh. But the majority of the pilgrims take the pathway from Nandya which is highly hazardous journey but in consonance with traditional philosophy of salvation through penance. Pachmarhi draws pilgrims from the Vidarbha region of Maharashtra. They travel by bus to Chhindwara or Tamya and then start walking. Some take Dasmoh route while some take Kuabadla route. Finally the groups reach the shrine of Bhura Bhagat at village Sengakheda. After seeking blessings from the great devotee of lord Shiva Bhura they proceed for Nandya.

Both at Sangakheda and Nandya camping facilities are available on rent at the rate of Rs.5 per person per night. They make offering to Nandya deo. Some also reach Nandigarh on a hillock where stone image of Nandi the divine bull is there. They come down to Nandya and then take up the actual hazardous trail to reach Chauragarh. The path way is steep and at places along the ridges and any false step may lead to an accident.

The cave temple and hill top temple of Lord Shiva have made Pachmarhi a very sacred hill range. Each of these temples was under the direct management of Mawasi/Korku chiefs of the state of Dubka and Jhont. There were no priests of Brahmanic order. Pilgrims used to make offerings by their own. Some Korku used to assist and maintain the temples. Mahadeo cave temple management has been taken over by Dasnami Sanyasi Akhada. The Akhada has its own establishment adjacent to the cave temple. Members of the Akhada remain in the temple by rotation and help the pilgrims in making offering. At this juncture it is to be mentioned that in most of the Shiva temples the devotee make offerings by their own and the priest media is not essential. Some Korku Bhagats used to assist the devotee in making offering. At Bara Mahadeo it is being done by Akhada people round the year. The offerings are taken by them. The same Akhada manages the Chauragarh temple daily except the Shivratri fair (one week long). Chadotri (offerings) other than Shivratri are taken by the Akhada. Management is under the local administration. They used to auction the offerings which accumulate during the fair. There was a dispute with the Akhada and following court order from Madhya Pradesh High Court the local administration itself manage the affairs. The sub-divisional officer, Pachmarhi supervises it.

Traditionally both the temples belonged to Korku and managed by the representative of the Korku rulers. So Korku who refer themselves as Mwasi Thakur feel attached to this temple more than the Gond the largest tribal group in the region. The Mawasi Korku spread stretches from Chhindwara Hoshangabad in the east to Khandwa and parts of Dhar on the west. Mainly they are distributed in Mahakoshal area Western Part of Pachmarhi bio-reserve there are Mawasi/Korku. From eastern part Gond area commences. The Bharia have a good population near Tenya,

Patalk of Chhindwara district and adjacent areas of Balaghat district. Gond group of tribes and Bharia or Bharti live side by side and even in same village but different hamlets. In the study villages Mawasi/Korku take special interest in Shivratri yatra. Gond and Bharti do not always bother to have pilgrimage to Mahadeo on these occasions. Territorially the Korku feel it is their area as they belong to the ruler's group. This is why they prefer to be called as Mawasi Thakur or Thakur Korku a status group in itself.

Mahashivratri fair continues for eight days or a week and falls in March. Rakshabandhan and Nagpanchami in August and Dev Uthni Egaras in October. The main pilgrim flow is in March on the occasion of Shivratri in which our cluster of village gets involved. The villages described are revenue villages and agricultural farming is the main stay of their economy. Collection of forest products and daily wage in construction or forest work are to supplement their earnings. The fair provides enormous opportunity to earn cash and accumulate. Lord Shiva's association with liquor and other intoxicants, allows the devotees to drink. There is a strong tradition of offering chicken and goat and metal trident. The pilgrims bring trident of odd kilogram of weight from their nearby towns and leave the same at the temple of Chaurababa as offering towards fulfillment of a vow taken. Around two truck loads of trident in the temple premises speaks of this tradition. The other common offering is that of poultry bird and goat which increases the demand several times. So the villagers raise chicken and goat in a way these will be fully grown and ready for offering at the time of Shivratri fair. Firewood, poles, logs, bamboo are also in great demand in the fair. The pilgrims and also the shopkeepers require these. The commodities stated here give them a good return. They distill the Mahua liquor and every family make a stock of minimum 500 litres. This gets exhausted in the fair. In true sense any amount of stock of Mahua liquor is likely to be sold. As they can not invest more so they prepare that much of liquor.

It is believed that only under the spell of liquor the pilgrims scale the height along the risky pathway. They also open shops of eatables, special soil for plastering in different colours, and so on. Young boys earn by serving as porters or guide or helper. In a day each earns around two hundred rupees or more. This is the sole cause that the villagers of Nandya, Gutkheda, Binora, Digri and Sukhdongar do not go to the plains for Ravi harvest. They just cannot ignore this source of earning.

The Panchayat, Health, Forest, Irrigation, PWD and Food & Supply departments take this occasion specially as a challenge. Nandya Panchayat acts as a nodal agency. Shop sites are distributed on payment of charges fixed by the Panchayat. (Rs. 200/- for small, Rs. 400/- for medium and Rs.800/- for large hotels). Parking place, barriers are auctioned. Priority was given to local bidders this year. Korku of Gutkheda won the bid. It was decided to make available drinking water round the clock through tankers. The anicut in the river is closed to store water. A stock of bamboo and wooden logs made ready to provide to the shopkeepers on rent. Health department opened the dispensary with emergency services. One ambulance was kept ready. The medical team camped at the village school. Lot of accidents take place keeping the medical team busy round the clock. Pilgrim flow remains during day as well as night hours. Food and supply department raise special quota of rice, wheat, sugar and kerosene oil to make available through fair price shop. The forest check post make a stock of fire wood logs depot to make it available to the shopkeepers. Additional staff is mobilized to prevent damage to the forest by the shopkeepers and pilgrims. The department took up awareness programme for the pilgrim tourist about forest resource and threat it is being subjected to. This was attempted by fixing posters, festoons and distributing handbills. The department is using this occasion as an opportunity. Necessary repairing of fair weather roads were taken up ahead of the onset of this fair.

The Mawasi have a number of exogamous totemic clans. Tribal endogamy is strictly followed. Of the three tribal groups living in our cluster of study villages Korku/Mawasi being the rulers' tribe claim highest status followed by Gond. The Gond here has one section but in the area there are other Gond groups like Pardhan, Ahir etc. Pardhan were the priest of the Gond- a status which they have lost as the Gonds are taking service of the Brahman (low category) priest.

Table 3.21

Tribe-wise Distribution of Family by Types

Family Type	*Tribe*							
	Korku/Mawasi Thakur	*%*	*Gond*	*%*	*Bharti/Bharia*	*%*	*Total*	*%*
Nuclear	83	53.54	18	43.90	12	54.54	113	51.83
Vertically Extended	44	28.38	11	26.82	8	36.36	63	28.89
Horizontally Extended	6	3.87	7	17.07	0	0.00	13	5.96
Mixed Extended	5	3.22	2	4.87	0	0.00	7	3.21
Broken	17	10.96	3	7.31	2	9.09	22	10.09
Total	**155**	**99.97**	**41**	**99.97**	**22**	**99.99**	**218**	**99.98**
Average Family Size	**5.03**		**6.8**		**5.8**		**5.9**	

The tribal groups under study have a preponderance of nuclear type of family. But all tribes do not stand at the same plane as is evident from the presence of extended type of family among the Gonds in more number as compared to that of Korku. Table 3.22 depicts tribe-wise distribution of family types as well as average size of the family.

This speaks of the dependency of the Gonds on agriculture. The average family size among the Gonds is not only higher than that of both Korku and Bharti but also than the common average size supports this view. In the gala affair of Mahashivratri fair the Gond have a marginal role to play. Economically it has a partial significance.

Both Mawasi and Gond claim to be Hindu, practice Hinduism alongwith their tribal animistic belief system. The Korku are more Hinduized in this regard. Both of these groups desist from taking beef. The Bharti or Bharian enjoy the lowest position in the local hierarchy as they are believed to take beef. There is a case of Korku – Bharia marriage. The Korku boy has been thrown out of the Korku society though Bharia has accepted the intertribal union and recognized the marriage. Korku do not accept food and water from Bharti.

Village exogamy is the general rule but the average marriage distance is fifteen kilometres or even less. Payment of bride price is mandatory among them. The poor boy has to serve his would be father-in-law for five years to secure a bride. The contract is finalised at the panchayat. In these villages except in case of Nandya village panchayat is also the ethnic panchayat. The serving groom is called Lamjhana and the system as lamijhana marriage. Widow re-marriages are in practice. Younger unmarried brother of the deceased retains the claim on his elder brother's widow. In such cases the second husband need not pay bride price as it was paid by his deceased brother. In case if there be no unmarried younger brother of the deceased of more or less matching age or if he declines to marry his brother's widow, the lady is at liberty to go back to her father and marry again. Then also the bride price is paid by the second husband to the widow's father. Divorce is permitted as per customs as well as the divorce remarriages. The second has to compensate the first husband by paying the bride price. In case of elopement the union is regularized after payment of higher bride price. Under any circumstances clan exogamy has to be observed. Marriage by intrusion is rare but possible. If both the parties agree union is recognized and then girl's father claims low amount of bride price. A widower can marry his deceased wife's younger sister by paying full amount of bride price. So a Korku marriage involves money. This is also true for other life cycle rituals. Death rituals particularly Sidoli involves lot of expenditure. This involves planting of memorial posts at the clan ancestor's place. It continues for three days. This they perform in the month of March and April.

As per the local tradition marriages can be solemnized after Deepavali. But they perform marriages after Shivratri fair. More than two months before Shivratri they get busy in making preparations like distilling Mahua liquor, collecting fire wood, splitting and cutting into thinner pieces etc. They cannot afford to spoil two-three days for any celebration. Again after Akshay Tritya agricultural work cycle commences. The period between Shivratri and Akshay Tritya is the peak marriage season. A last mortuary rite for all deceased members of the lineage during the preceding one year is also performed mainly during this period. Kharif produces are mainly for their own consumption. A small portion of the Ravi crops grown after Deepavali they use to sell for cash. Also they earn from selling minor forest products collected around this period. Finally earnings from the fair enable them to meet the extravagant expenses for mortuary rites and performances of marriage. After they become free following the conclusion of the fair and have some good amount they plan for solemnization of marriage etc. The girls start singing marriage songs in the night and dance to the tune of the drum. Gatha or Sidoli the last mortuary rite continues for two to three days. All able persons take part in the marathon dance.

After the marriage season the work cycle for the kharif crop starts. Then comes Rakshabandhan fair which is a day long affair. People go to Mahadev cave temple tie Rakhi and seek support from the lord. They make offerings as per the vow taken last year. Chicken and goat offering is the important part of it. In couple of months the chickens become grown up and ready for sacrifice.

The fair is a day long affair and the pilgrims are from adjacent districts. The next fair held on Dev Uthni Egaras which marks the end of four month traditional slumbering period. During this period activities centre round work cycle of Kharif crop. Also it is the rainy season which restricts the movement to distant areas. Celebrations remain suspended. It is believed that deities retire during this period, saints also camp at a place. Dev Uthni Egaras is celebrated as the beginning of celebrations. In this day long fair there remains demand for chicken, goat and liquor. In this occasion the pilgrims from adjacent districts come and pass through this village on their way to Bara Mahadeo.

Shivratri fair continues for eight days and pilgrims come from other states also. Majority come from Vidarbha region of Maharashtra. One satsang society (Nandya kheda Seva Samitee) having its headquarter at Itwasi area of Nagpur has a Dharmshala here to cater to the needs of the pilgrims from Nagpur. Another Dharmshala is under construction. Pilgrims also come from the districts of Mahakoshal area. In terms of number of pilgrims and stretch shivratri is most important among all the three. Of late shop owners also sell poultry birds. But the devotees always prefer chicken raised by the tribals even at more than double of normal rate.

Shivratri is most important festival for the Mawasi/Korku. Due to this fair they cannot have darshan of lord Shiva on the day of Shivaratri. They will be busy in economic activities. To have darshan of Chaurababa and Bara Mahadeo is equally important. They visit the temples before the onset of the fair and then get set to their work. After the completion of Shivratri fair the village, mainly the main road which is the route of the pilgrim is cleaned. Everybody plasters his house as they did before the fair. Additional structures erected for providing sleeping speak to the pilgrims on rent is demolished. Many aged pilgrims come who fall sick and even expire. Every year there are fall cases, dehydration etc. It is necessary to clean the village from the dirt accumulated. After cleaning the individual house and common places in the village worship of Hardulala is arranged. Each household contributes and the Parihar or Bhagat or priest make the offering to the deity. Individual as well as communal offerings are made. Importance is attached to village level offering. Hardulala is worshipped after Rakshabandhan as well as Dev-Uthni-Egaras too.

Bidri is other important festival observed in the month of Jeth (May). After this sowing of Kharif crops starts. In the month of Savan (July-Aug) Kharreri Deo is worshipped. Coconut is offered and worship is performed in farm land. Kodo and Kutki which is short term kharif crop to be harvested are offered to Kangri, Makka and Jawar. Kodo Kanghi and Kutki harvest is in Kartik (Octo-Nov.). Sawa is the first crop harvested in the month of Bhado (Aug.) Bhado Amavas (new moon) is celebrated as Pola. It is a festival of cattle. They become free from agricultural operation. The cattle are taken around and fed green grass. A month earlier unmarried girls make a couple out of mud. Every evening they sing songs before this doll couple which continues for the whole month. On the Amavas day the doll couple is taken in a procession of the unmarried girls with songs, dance and drum beats to river Denwa where the divine couple is immersed. The divine couple is worshipped as lord Shiva and his consort. On the fifth day after this is observed as Nagpanchami.

Navratri they observe in the month of Kuar (Oct.) They keep fast (single meal) on all the nine days and on the Navami (9th) day offer coconut to goddess Durga. This is followed by Deepawali when Laxmi pujan and Gaumata (cow) pujan is performed. Makar Sankrant is an important festival observed on the last day of the month of Pus (Jan.). They make offering to Anhoni mata at village Jhirpa. Offerings are made to dead ancestors. This is called Radhoni which continues for one month.

The people of the village cluster under study have a kind of orientation to the flow of pilgrim tourists in each of the three fairs. The most important fair in all respects is that of Shivratri and the Korku are the main tribal groups linked up. Politically and territorially they are attached to Lord Mahadeo. Their and also other tribals economy is under the influence of this pilgrim influx. They have oriented their socio-ritual aspects to accommodate the fair. Symbolically they are attached to lord Mahadev and this fair. They make necessary adjustment not to hamper, their schedule during the fair. Delving in the stories of Lord Shiva they get involved in their economic pursuits. Traditional tourism as depicted here the local tribal acquaint them to the process so that it should run smoothly. They offer resting place, drinking water to the pilgrims. Also they take care that the pilgrims should not be cheated. One such case was brought to the notice of the author where the thief was severely beaten by the group of boys who look after the safety and security of the pilgrims. The process is part of life of the local tribal groups in general and Korku in particular.

4

The Watershed in the Core Zone Villages of the Pachmarhi Biosphere Reserve

Surajit Kumar Pal

Water, one of the free gifts of nature, has been sustaining most of the species and the food chain on the planet, since the beginning of creation. Without it all life ceases. That is why it is considered **light most vital and basic component of life.** Human beings and all other living things i.e. the animals, birds, insects, plants and various species exist due to benevolence of water. Historically proved that water sources determine various human settlements, civilization, trade and migration routes. It influences the entire life system, ecology and environment of the earth. Hence the famous German poet Johann Wolfgang Von Goethe writes:

> "Everything originated in the water
> Everything is sustained by water"

This unique gift, exclusive probably to this planet, should be sustained through pragmatism and made best use of, without either polluting or wasting or overdrawing it. As it is the basic and most vital component of life and has direct role in maintaining the life system of the earth, its sustainable use has not been followed by the masses of the people who are mostly profligate in nature. It calls for an integrated and scientific approach supplemented by prudence and restraint from its abuse. Rain and both underground and surface water sources should be enriched using human ingenuity. People should not resort to actions that are detrimental but do everything to enrich perennial character of rivers, lakes, streams, tanks etc.

The increasing human population, bovine and other sub-human species pose larger demand for water. This increasing biotic pressure leads to over exploitation of such prime natural resource that also causes the depletion of the non-perennial water sources, erode underground water sources and pose additional strain on perennial water sources. Enlarged area has become inevitable to meet the food and fodder needs of increasing human and animal population. Under this situation people should stress on improving yield per hectare rather than enlarging area under irrigation through less water-intensive crops and adoption of scientific water management techniques.

Every drop of water should be judiciously utilised and its reckless wastage eliminated. Not all regions of India are endowed with water sources. There are several zones that are facing severe water crisis, while some other parts of the country are endowed with surpluse water which goes into sea without proper use. To overcome acute water crisis and also to maintain natural balance, watershed programme is gaining momentum throughout the country. This is not a new concept as in the ancient time the tanks, canals, wells and ponds were built for similar purposes. Under this programme existing tanks, ponds, wells in every village which may be abandoned or silted should be restored for rain water harvesting for meeting increasing water needs and also to develop watershed by making new ponds, tanks and wells and afforestation methods etc.

During the ancient periods digging of tanks, canals, wells, *barawas* (small dam) etc, was considered to be a pious work and remnants of these water resources are still found in different parts of the country. Development and management of these resources were people oriented i.e. people used to take care of these resources for maintenance and development, also deeply embedded in the cultural norms of the Indian society. But in course of time particularly during the British era it was made govt. oriented that still continues even in independent India. The importance of earlier system of management of the water resources needs to be understood as well as to be restored at this stage when the whole world is facing severe water crisis. Now watershed has also become recognised as a key and direct development unit in rural development.

The term watershed implies that the area of land from which precipitation flows to a single stream or water body. It is basically a geo-hydrological unit with a common drainage outlet. It is synonymous as 'catchment area' and 'drainage basin'. In other sense, it is a land surface bounded by a drainage divide which contributes runoff to a common drainage point. A watershed may be a few acres in area or may consist of a drainage basin of a larger area in size, which is demarcated by ridges or gullies. In real sense, a watershed is made up of soil, vegetation and water along with the people and animals who are the integral part of system. Watershed, drainage area or catchment area is generally interchangeably used meaning for the same.

In recent time the term watershed has become a subject of discussion. Because the people now have come to realise that major part of their well-being relates to a watershed in some manner or other. The water which we use comes from a watershed; and destructive floods also come from a watershed. Thus, watershed has two broad characteristic features - destructive and constructive. From destructive standpoint it creates havoc to the people in the form of flood that destroys the habitats of plants and animals, human settlements, crops, etc. While in constructive standpoint it has several beneficial aspects such as hydroelectric power generation, navigation, irrigational water supply, municipal and other local bodies' water supply, waste disposal etc.

Watershed deals with three major natural resources - land, forest/vegetation and water. These three resources are closely interwoven that influence the perennial character of watershed and also the eco-system of an area. Development and management of land, conservation of forest and water resource are the essential components of watershed. Land erosion and deforestation lead to deplete the groundwater recharge as well as biomass. Intensive agricultural practices, industrialisation and urbanisation for which groundwater and surface water are excessively exploited that are largely responsible for damaging the eco-system of the earth. As water supports the whole life system its judicious use needs to be adopted at this juncture when the whole world is facing severe water crisis. The present day water crisis is the resultant effect of man's irresponsible and selfish attitude towards over exploitation of natural resources leaving aside flora, fauna and other animate and inanimate objects of the earth whose contributions are immense in maintaining ecological balance. In dealing with the watershed management its three major components (land, forest and water) are required to be examined.

Under this paradigm an attempt has been made to focus the existing status of the watersheds found in the core zone villages of the Pachmarhi Biosphere Reserve (PBR), Madhya Pradesh that may not only enrich our knowledge about the system of conservation and maintenance as nurtured by the people generation after generation but also to maintaining the biodiversity of the area which are being depleted and contaminated by constant human interventions in respect of agriculture, grazing, tourism, human settlement, forest based enterprises, etc.

Pachmarhi Biosphere Reserve (PBR):

The PBR declared on 3rd March 1999 by the Govt. of India, lies between latitude 22° 112 to 22° 562 N and 77° 472 to 78° 522 E longitude, covering an area of 4987.38 sq. km. of which 59.54 per cent falls under the district of Hoshangabad, followed by Chhindwara(29.20 per cent) and Betul (11.26 per cent) districts of Madhya Pradesh. Within its jurisdiction three wildlife conservation units viz. Bori Sanctuary (485.72 sq. km.), Pachmarhi Sanctuary (417.78 sq. km.) and Satpura National Park (524.37 sq. km.) are located that also constitute the Satpura Tiger Reserve (1427.00 sq. km.), notified in August 2000. Out of its total area, the entire Satpura National Park is regarded as the core zone and the remaining area of 4525.93 sq. km. is considered the buffer zone.

According to the official record its bulk part (62.47 per cent) is under forest cover and the remaining part (37.53 per cent) is under cultivation (29.38 per cent), water bodies (4.94 per cent) and waste land (2.18 per cent), also having a built up area (0.54 per cent), containing 6 urban and 574 rural (Revenue village - 507 and Forest village - 67) settlements, distributed under ten tahsils of the aforementioned three districts.

As its large part falls under the Hoshangabad district, also considered a district of rivers, its importance in terms of richness of bio-diversity, river system, minerals etc., is immense not only to keeping ecological balance of the vast area but also to the people of the area for their economic enterprises. Physiographically the district of Hoshangabad is broadly divided into two parts - the hilly part, consists of famous 'Satpura Hill Range', covering the whole of the south, rising in a continuous chain of forest-clad hills, highly dissected by the Tawa and other tributaries of the Narmada and the valley covering the central and northern parts of the district. The PBR also covers primarily the hilly parts of Betul and Chhindwara districts. Thus, rivers, valleys and hill ranges contribute to the ecological riches of different flora and fauna in the PBR.

Fieldwork and the Study area

The study area is confined to eight core zone villages - Bariaam, Dhana Pagara, Badkachhar, Kajri, Rorighat, Binora, Chitor Pattan and Godanar located within the boundary of the Satpura National Park (SNP) of the PBR. Of all villages first seven are of revenue category and the remaining one is forest type. The fieldwork was conducted for one month duration, commenced from 9th February to 9th March 2005 including the journey period in order to gather first-hand information about the status of watershed/ water resources prevailed in the core villages on which the sustenance of the people, vegetation and other living things depend. Having been short stay in each village (average field visit per village was 2-3 days) proper anthropological methodology could not be followed in culling the field data. Considering time constraints as well as coverage area (i.e. nature of geographical location of the study villages) finding no other alternative way survey method intermingled with some sort of observation was ultimately adopted to gather the preliminary information on the concerned topic.

Barring Bariaam all other villages are located in difficult terrain, not easy to access, surrounded by forest. The forest roads which are jeepable after monsoon lead to approach the villages, inhabited predominantly by the Adibasi population - Gond (Bariaam & Chitor Pattan) and the Mawasi (Dhana Pagara, Badkachhar, Kajri, Rorighat, Binora and Ghodanar villages). Bariaam is located along the Pachmarhi - Piparia road, about 7 km. away from the Pachmarhi town. Average distance from Pachmarhi to all other villages except Binora and Chitor Pattan is about 15 - 20 km. Binora and Pattan are located over 100 kms. away from Pachmarhi. Inaccessibility has further been restricted by imposition of ban to enter into the STR. Hence no outside person can either enter or stay in the any core zone village without the permission of the STR/PBR authorities.

The population distribution by sex and literacy as per 2001 census reveals that the male-female ratio does not bear much difference and the average household size is around six. But majority of the population are illiterate (75.00 per cent), only 25.00 per cent population are the literates of which the female literacy is very unimpressive (8.00 per cent). And most of the people who are considered as literates can put their signature only.

It is not properly known how the tribal villages came into being in the forests. However, the historical records indicate that the villages found in different parts of the old Pachmarhi and Bori reserve forests (now constituent part of the PBR) are the resultant effect of existence of different jagirs set up in the pre-British era. In Hoshangabad district several jagirs were in existence of which Chhatar, Bariaam-Pagara and Pachmarhi jagirs are notable, held by the Gond and the Korku Rajas or Chiefs. And the jagir system led to set up various tribal villages inside the forests as also seen today.

Occupationally the people of the villages are, by and large, agriculturists and the land they own is inadequate in maintaining their livelihood. Majority of them are primarily the marginal farmers-cum-owner-cultivators and at the same time they are the wage labourers as their agricultural produces are insufficient to maintain their sustenance throughout the year. Landlessness also prevails in the area. As they are, by nature, the forest dwelling communities, whose subsistence largely depends on harnessing of available natural resources and agricultural activities, both male and female take part in different economic activities such as collection of forest produces(fuel wood, fruits, house-making materials - grasses, leaves, ropes. wooden posts etc.), agricultural and other menial works.

Land system:

In dealing with watershed a brief history of land system has also to be mentioned as it bears some relevance in the present context. Because without understanding the history of land tenure/ management any kind of watershed programme cannot be undertaken or executed. The constituent districts of the PBR have had long tenural history of which Hoshangabad part is more significant. The district of Hoshangabad was part of the Gond dominion and no proper records showing what sort of land revenue systems existed during their regime. However, some factual records envisage that the district was divided into talukas held by the jagirdars, who sub-leased the villages at their own pleasure to the cultivating classes of the population. When the Marathas extended their conquests to this district, Hoshangabad and its adjacent areas were in the hands of the Mohammedans of Bhopal and the rest of the district including the hilly part was ruled by the chiefs of Gond and Korku tribes whose chiefships were not destroyed by the Marathas (Corbett & Russell 1908:227). During the British regime the jagir system remained undisturbed.

While in the valley area the Marathas had the stronger hold, though the Gond Rajas of Fatehpur and Sobhapur (both in Sohagpur tahsil) were continued in possession of their country, their powers of sovereignty were strictly limited. The tribal chieftains occupying the Satpura hills around Pachmarhi were also allowed to remain in their jungles as a service tenure holders, under which they were required to keep the pathways over the hills from the marauders, so as to permit pilgrims to visit the sacred shrine of Mahadeo in the Pachmarhi hills. Lower down the valley the Korku headmen of Damjipura and Bairi retained their estates, but in the rest of the district the aboriginal leaders were shorn of their possessions. The Marathas did not create any large estates in their place, except in a very few instances as muafi or revenue-free grants for distinguished services.

The land revenue system during the British rule was reshaped with various settlement operations from time to time in order to gather revenue for consolidating their position in this part of the country. As part of forest management during their administration forest villages came into existence on forest lands with the purpose of uninterrupted labour supply for extraction of timber, making of forest paths, regeneration works etc, as found in other parts of India. The people living in forest villages located within the reserved forests were supposed to do the forest related works under some terms and conditions without any wage. Instead the reclaimed forest patches given to them for making dwelling units as well as for cultivation, were devoid of any ownership right, familiar as *tangya* system prevailed in Bengal and Assam. On the contrary, during British rule in central India both forest and revenue villages were in existence within the very boundary of the forests that still prevails.

In post-independence era, particularly after declaration of the national park, sanctuaries and then biosphere reserve the status of the actual tillers of the land living in both revenue and forest villages located under the PBR has become more complicated, rather obscured on the issue of ownership right on land under their possession. In fact, now the land tenure of the PBR area in the hands of two authorities - Revenue & Forest Depts. This dual authorityship creates problems in executing different development programmes meant for the people and also for conservation and management of forests which are important in keeping the perennial and longevity character of different watersheds spread in the area.

Cropping pattern

The cropping pattern is an important dimension of watershed management. It determines largely the degree of water requirement of an area. On the basis of climatic/weather condition and available soil and water resources the people grow various crops season after season and year after year that in time perspective such crop varieties are considered as the cropping pattern to the people of that particular area. And nature and technology of production are devised according to soil types and availability of water resources.

Table 4.1
Cropping Pattern in Core Zone Villages of the Pachmarhi Biosphere Reserve

A	*Kharif crops :*	*Name of the crop*	*Local term*	*Sowing season*	*Harvesting season*
I	**Cereal crops**	Kutki	Kutki	May-June	Sept.-Oct.
		Kodo	Kodo	June-July	do
		Sama	Sama	do	Oct.-Nov.
		Sorghum	Jowari	do	Nov.-Dec.
		Bulrush millet	Bajra	do	do
		Maize	Makka	May-June	Aug.-Sept.
		Paddy	Dhan	June-July	Nov.-Dec.
II	**Pulses**	Pigeon pea	Arhar or Rahad	May - June	Feb - Mar.
III	**Oil seeds**	?	?	?	?
IV	**Vegetables**	Gourd	Louki	May - June	Aug.-Sept.
		Pumpkin	Kaddu	do	do
		Brinjal	Baingan	do	do
		Lady's finger	Bhindi	do	do
		Chilli	Mirchi	do	do
		Cucumber	Kakdi	do	do
		Turai	Gilki/Turai	do	do
		Kidney bean	Chaulai/Barbati	do	do
		Kidney bean	Ballhar	June-July	Jan.-Feb.

(Contd.)

(Contd. Table 4.1)

A	*Kharif crops :*	*Name of the crop*	*Local term*	*Sowing season*	*Harvesting season*
B	**Rabi crops:**				
I	**Cereal crops**	Wheat	Pisi	Oct.-Nov.	Mar.-April
II	**Pulses**	Gram	Chana	do	Mar.-April
		Pea	Matar	do	Feb.-Mar.
		Moong	Moong	Dec.-Jan.	April-May
III	**Oil seeds**	Mustard	Sorsu	Oct.-Nov.	Feb.-Mar.
IV	**Vegetables**	Tomato	Tomatos	Oct.-Nov.	Feb.-Mar.
		Potato	Alu	Oct.-Nov.	do
C	**Horticultural/**	Mango	Aam	-	-
	Fruit crops :	Tamarind	Imli	-	-
		Jack fruit	Kanthal	-	-
		Guava	Amrud/Jam	-	-
		Banana	Kela	-	-
		Papaya	Papita	-	-
		Plum	Ber	-	-

N.B.:-1. Kutki, kodo, sama & jowar used to be considered the staple crops of the forest dwellers till the recent past, almost abandoned now.

2. Cultivation of rabi crops found in different core zone villages have been introduced in recent years.

3. Barring mango & plum (which are naturally found in both forests & villages) all other horticultural/fruit crops mentioned in the table are being grown recently in several core Zone villages.

In the PBR area where both tribal and non-tribal people live, crops are grown mainly in two seasons - *khairf* and *rabi* and the crop varieties produced in both seasons are known as kharif and rabi crops respectively. The crops grown in both seasons are normally categorised into cereals, pulses, oil seeds and vegetables. However, some horticultural or fruit crops are also seen to grow that may also be considered as part of the cropping pattern of the said area (Table 4.1).

Similarity exists in respect of cropping pattern found in all study villages (Table 4.2) except Bariaam where practice of agriculture is absent. Wheat cultivation recently gets more importance now to the villagers than any other cereal crops as some sort of irrigation facilities are available, though not in absolute terms, specially after construction of stop dams in some perennial springs/streams flowing nearby some core zone villages. This (wheat) has become their staple crop now.

Cultivation of kutki, kodo, sama, jowar crops, used to be considered their staple crops, grown under rainfed condition in the hilly slopes when *dhaya* (slash-burn type) cultivation was in vogue, has reduced drastically due to imposition of ban on slash - burn agriculture and various other restrictions under biosphere rules. Only a few villagers produce these crops in a very rudimentary manner in their land, but not by slash-burn method. Now the people (rather Adivasi cultivators) of core zone villages have abandoned the practice of slash-burn agriculture, and have switched over to settled cultivation. Maize cultivation is commonly seen in all villages, usually done in small patches that may be their courtyards/kitchen gardens or some fallow land located adjacent to their residential places.

The cropping pattern and technology of production, now in practice, is the resultant effect of various restrictions imposed by the forest/biosphere authorities, in one hand and on the other their constant interactions with the non-tribal peasants living in buffer and outside the buffer areas. Method of cultivation now used for producing different crops in study villages do not segregate them (Adivasi peasants) from the non-tribal/caste peasants as far as technology of production and production organisation are concerned. However, production relations are yet to be entering complex structures as markedly observed in the non-tribal peasant communities.

Table 4.2
Village wise Cropping Pattern

Sl.No.	*Name of the Village*	*Crop Varieties*				
		Cereal Crops	Pulses	Oil seeds	Vegetables	Horticul/ Fruit crops
1.	Bariaam	-	-	-	Gourd	Papaya
					Kidney bean	
					Tomato	
2.	Dhana Pagara	Jowar	Pea	Mustard	Gourd	Mango
		Maize			Pumpkin	Plum
		Paddy			F. bean	Jack fruit
		Wheat			Brinjal	Tamarind
					Kidney bean	Guava
					Turai	Banana
					Cucumber	Papaya
					Chilli	
					Tomato	
3.	Badkachhar	Jowar	Pigeon pea	Mustard?	Gourd	Mango
		Maize			Barbati	Jack fruit

(Contd.)

(Contd. Table 4.2)

Sl.No.	Name of the Village	Crop Varieties				
		Cereal Crops	Pulses	Oil seeds	Vegetables	Horticul./ Fruit crops
		Paddy			Kidney bean	Guava
		Wheat			Potato	
4.	Kajri	Kutki	Gram	?	Gourd	Mango
		Kodo			Brinjal	
		Jowar			Kidney bean	
		Maize				
		Paddy				
		Wheat				
5.	Rorighat	Kutki	Gram	Mustard	Gourd	Mango
		Jowar			Pumpkin	Guava
		Maize			Kidney bean	Jack fruit
		Wheat				
6.	Binora	Kutki	-	-	Gourd	Mango
		Kodo			Pumpkin	Jack fruit
		Maize			Kidney bean	

(Contd. Table 4.2)

Sl.No.	*Name of the Village*	*Crop Varieties*				
		Cereal Crops	Pulses	Oil seeds	Vegetables	Horticul./ Fruit crops
		Wheat				
7.	Chitor Pattan	Kutki	Pigeon pea	Mustard	Gourd	Mango
		Kodo	Gram		Pumpkin	Plum
		Sama	Moong		Brinjal	Tamarind
		Maize			Kidney bean	
		Paddy				
		Wheat				
8.	Ghodanar	Jowar	Pea	Mustard	Gourd	Mango
		Maize			Pumpkin	Jack fruit
		Bajra			Brinjal	
		Paddy			Chilli	
		Wheat			Lady's finger	
					Tomato	
					Potato	

Water channelised from stop dams for irrigating the crop field during rabi season does not serve the interests of all the cultivators as the command areas are limited in nature. This leads to some kind of conflicts being subtly brewed among the deprived fellow villagers, though not overtly exposed, in respect of sharing the water for irrigating their land. The watersheds located in majority of study villages are of micro or mini type, cover only a limited area of land. In spite of this wastage of water flowing from different streams (still unregulated) is a common feature in the area that needs to be arrested by modern devices in order to make more area arable, so that the people can improve their economic condition vis-à-vis quality of life to some extent. In fact, water security needs to be stressed through which food security of the people can be ensured in the long run i.e. both food and water securities are interwoven and also inseparable in character.

Soil Condition

Soil condition is one of the vital dimensions of watershed, linked closely with its (watershed) management and development. Percolation of rainwater into earth that contributes to groundwater recharge and moisture retaining capacity of soil largely depend on soil types. Soils of the PBR (i.e. three constituent districts) are officially classified into two broad categories - (i) shallow & medium black (Chhindwara and Betul districts) and (ii) deep-medium black (Hoshangabad and other districts), based on physical nature and fertility condition of the soil. Despite such classification the people living in the study villages have their own connotations on soil types, perceived by them generation after generation.

According to them, soils are divided into three types, namely *reteli* (sandy), *domatia* (sandy loam) and *kapo* (clay mixed alluvial). All these soil types barring *kapo* variety are less fertile and having little water retaining capacity. Crops like *kutki* (Panicum miliare), *kodo* (Paspalum scrobiculatum), sama, maize and some vegetable varieties like gourd, pumpkin etc, are grown. The kapo type is fertile, black in colour, possessing the quality of more water retaining capacity than other categories due to its very fine texture. The kapo type of soil is suitable for wheat, rice, pulses and vegetables.

The undulating hill slopes are having mainly the reteli type of soil. Soils found in the villages of Bariaam, Dhana Pagara,

Badkachhar, Kajri, Rorighat, Binora and Ghodanar are of *reteli* (sandy) type. While domatia and kapo types of soil prevail in Chitor Pattan village (Table 4.3).

Table 4.3
Soil Types in Study Villages under the Pachmarhi Biosphere Reserve

Sl.No.	*Name of the village*	*Type of soil*	
		Local term	**English term**
1.	Bariaam	Reteli sera	Sandy
2.	Dhana Pagara	do	do
3.	Badkachhar	do	do
4.	Kajri	do	do
5.	Rorighat	do	do
6.	Binora	do	do
7.	Chitor Pattan	a) Kapo, b) Domatia	a) Clay mixed alluvial b) Sandy loam
8	Ghodanar	Reteli	sandy

In this connection it may be relevant to focus the rock formations that bear some significance in understanding the nature of water retention and percolating capacities which prevailed in different soil types on which forest vegetations, agricultural practices, plant growth, condition of groundwater recharge etc. depend. Geologically the area consists of ancient *archaean* formations to the most recent alluvium and soils. The rock formations may be arranged in chronological sequence from which various soil types derived or contributed to form soil categories that found in the PBR and its adjacent areas (Table 4.4).

The table shows that from which series various types of soils originated of which Gondwana and Mahadeva series are relevant to the present exercise. Soils derived from Gondwanas are normally sandy shallow supporting stunted tree growth, divided into lower and upper Gondwana series based on fossil contents. The lower Gondwana system lies along the edge of Narmada plains, Sukhtwa river and also in the North - west of Bagra along the Tawa river. The upper Gondwana is further divided into the Jabalpurs and the

Mahadeva series. The Mahadevas are also divided into Bagras, the Denwas and Pachmaris. Its composition is highly variable, the rocks consist of sand stones, clays, lime stones and conglomerates. Soils produced from the Bagra conglomerates are shallow and reddish in colour, fairly suitable for forest growth.

Table 4.4

Chronological Sequence of Rock Formations in the PBR

Sl. No.	*Era*	*Series*	*Types*
1.	Recent	-	Alluvium laterite soils
2.	Cretaceous-Eocent	Deccantrap	Silts, dykes, flows, inter-trappeam bed
3.	Permina-Jurassic	(a) Gondwanas (b) Mahadeva (c) amuda	(a) Upper Jabalpurs, sand stones and conglomerates (b) Bagra, Denwa and Pachmarhi (c) Lower Bijori and Talchirs
4.	Archaean	Metamorphic	Calcareous crystallines, solomites, phyllites and schists, granite gneiss and quartzites

The Denwas consists of alternating beds of sand stones and variegated clays, often greenish yellow and bright red mottled, often having *kankar* nodules, produce stiff coloured clay. It is found in the undulating valleys of the Denwa river and also in Anhoni-Amadeh tract. Pachmarhis are composed of sand stone, clays and platy veins of hard ferraginous matter. The soils derived from the Jabalpurs and Pachmarhis sand stones are primarily shallow and sandy, capable of supporting only stunted forest growth. Alluvium covers the entire Narmada plain, often yellowish to brownish with numerous inter-located bands of sand and gravel. Black cotton soils are fertile, derived from basaltic traps, granite and gneiss. Metamorphic rocks produce shallow soils capable of holding sufficient moisture, suitable for teak and saja forests. Soils derived from trap rock are more deep distributed along the river banks.

The lands available to the people of study villages are less fertile as the soils are mainly sandy and rocky in character which lack sufficient water retention capacity and the crop yields of different crop varieties seem to be comparatively less than the crop yields produced in sandy loam and clay mixed alluvial soils. The denuded hill slopes, rather the foot hills having some elevations where the crops are grown by the people, are devoid of humus contents as well as top soils, usually washed away due to surface runoff during monsoon. As a result stone pebbles/nodules with thin soil cover come out that do not hold adequate soil moisture, important for crop and other vegetational growth. When rainfall is of high intensity, only a portion of rainfall can enter the soil and stored in the root zone. In case of light rains of low intensity depending on the amount of moisture already present in the root zone of the crop, the entire rainfall may be effective rainfall. Effective rainfall is a part of rainfall available for the consumptive use of the crop.

The factors generally influencing effective rainfall are: amount and intensity of rainfall, consumptive use rate, moisture storage capacity of soil which depends on moisture holding capacity, initial moisture content and infiltration rate of the soil. As the villages located generally in the foot hills/hill slopes surface runoff is more and soil erosion is quite common in their area. No proper effective measures are adopted either by the villagers or by the Govt. agencies to arrest the soil erosion which is vital for crop production. Due to lack of required nutrients naturally available in the soil crop growth always remains stunted that affects agricultural production considerably. The nature of soil condition prevailing in the area prevents to hold adequate soil moisture during post monsoon period when the rate of evaporation increases due to scorching heat of the sun.

Forests

Forests have direct role in keeping the watershed alive perennially. Monsoon behaviour has positive correlation with land and forest ratio. Besides other advantages forests give birth to small streams and rivers and their perennial characters are sustained by forest cover and rainfall. Considering its importance in respect of maintaining ecological balance as well as biodiversity it needs to be protected. The constituent districts of the PBR have had rich forest cover (now degraded to great extent), divided into several types (Table 4.5).

Table 4.5

Forest Types Prevailed in the Pachmarhi Biosphere Reserve

Sl. No	*Name of the district*	*Forest types*
1.	Hoshangabad	(i) South Indian tropical moist deciduous slightly moist teak forests (ii) Southern tropical dry deciduous dry teak forests (with or without bamboo) (iii) Southern tropical dry deciduous mixed forests
2.	Chhindwara	(i) Southern tropical dry deciduous teak forests (ii) Southern tropical dry deciduous mixed forests (iii) Northern tropical dry deciduous peninsular sal forests
3.	Betul	(i) South Indian tropical moist deciduous slightly moist teak forests (ii)South Indian tropical moist deciduous mixed forests (iii)Southern tropical dry deciduous dry teak forests (with or without bamboo) (iv) Southern tropical dry deciduous mixed forests

Of the three districts, Hoshangabad covers the major part of the PBR (about 60.00 per cent), host to rich variety of wildlife, flora and fauna, and also very useful for maintaining ecological balance of the area. Forests of Hoshangabad district can be categorised broadly into two types - reserved and unreserved. This classification practically came into being in 1862 when the Forest Dept. was created first in India with a view to systematic conservation of forests! The reserved forests were directly under the control of the Forest Dept., while the unreserved forests were managed by the Revenue Dept. For smooth control as well as for proper management altogether six forest ranges - Bori, Sohagpur, Hoshangabad, Seoni, Rajabhorari and Kalibhit were created. The Kalibhit range, located in the extreme south - west of the district, bordering on the Tapti river was transferred to the Nimar district in 1904. The old Bori range, east of the Sonbhadra, regarded as the first forest recognised by the then British administration, was brought under systematic conservation in 1862 though in 1860 the first Conservator of Forests was appointed in Central Provinces. In 1862 the waste land rules were also introduced. This historical phenomena bears certain significance as far as conservation and management of forests are concerned.

Forest types mentioned in Table 4.5 are basically the deciduous category and play a pivotal role in keeping the perennial character of all watersheds spread in different parts of the PBR as well as the whole of Central India. Truly speaking forests of this region are watershed type and it takes hundreds of year to build up its own ecosystem, which has suffered severely due to unabated biotic interferences (primarily the humans) during the period of last two-three centuries. The end result is, it loses its carrying capacity in respect of watershed, soil fertility (particularly humus content in top layer of the soil), soil moisture, normal growth of vegetation, survivality of wildlife, and so on. In this context a brief discussion on the issue of deforestation and soil erosion may be highlighted in the following paragraphs.

Deforestation and soil erosion

Deforestation is a universal problem, that unfolded itself in the present era due to enormous population explosion and biotic pressure. The case of study area is no exception to this process. Historically Central India bears the witness of a long deforestation history started even prior to 1818 (this date is meant for the beginning of the British rule in this part of the country). Migration of different population groups and reclamation of forest lands for crop cultivation and other activities during medieval era and later period lead to cause enormous pressure on forests of the Gondwana region of the central highland India. The region was primarily inhabited by the Gond, considered the original population of the area who lived both in the plains and the hills, became outnumbered by the migrant population. In his book Captain J. Forsyth (1919) rightly mentions, 'the open country, under the rule of Akbar, was rapidly reclaimed by Hindu immigrants, arriving simultaneously from the north and the west... The Gonds retired to the higher plateaux and slopes of the central hills, where their hunting instincts, and rude system of raising the coarse grains on which they subsist, could still find scope' (p.9).

After the Mughals the region came under the Maratha regime which was ultimately defeated by the British only in 1818. In 1861 the Central Provinces were constituted as a unit of administration under the Chief Commissionership of Sir Richard Temple, of the Bengal Civil Service. He extensively toured the region of the Satpuras and found to his horror that unscrupulous contractors had already more or less taken over the forests and were felling them for the purpose of timber, especially railway sleepers. In particular Sir Richard Temple looked upon the region because of, "... its great capabilities for the storage of precious water..." (Forsyth 1919:14). He was thus the first person to understand the importance of these forests as protectors of the Great Central Indian Watershed (Buch, 2005:9). When the British rule was extended into this region, there began a race for occupation of unoccupied land with the objective of claiming title to it. While the contractors were busy decimating the forests, with no check on them because the forests were no man's land and at the same time there was the pressure of settlers trying to occupy more and more culturable waste land. This is one side of version related to deforestation of forests in central India. Undoubtedly the unscrupulous timber contractors were held largely responsible for damaging the virgin forests of this region, but during the British regime this was not fully arrested to control further damage despite the enactment of the Forest Act in 1927. Pachmarhi being the queen of Satpura and sanitarium attracted both the British officials and the Indian visitors that led to give the shape of a prominent tourist spot in later years and as a result substantial part of forests of the Satpura plateaus happened to be destroyed. And this caused partly depletion of the water level of the Great Central Indian Watershed as reflected in the sphere of rainfall, density of forest cover, siltation level of different existing rivers, streams, springs etc.

As Central India falls under the part of peninsular India its entire water regime is exclusively dependent upon the health of the watershed forests. Depletion of the watershed forests has taken place during the last over two-three centuries that has also affected the water regime largely as the rivers of this region are not fed by snowsmelt. The Narmada and the Tapti are the two principal rivers which flow westerly, considered the lifeline of central India specially to the people of Madhya Pradesh and its adjacent states. The life of these two rivers and their tributaries depends on the watershed forests of this region. The existing forests of the area cannot be treated as virgin as the same has been over exploited since the inception of state intervention on it. Thus, importance of conservation and protection of forests in this part of the country needs to be examined critically.

Deforestation and agricultural practices cause severe soil erosion in PBR area. Since the period of pre-colonial era, damage of forest vegetation in this part of the country started that still continues; but its degree of destruction in post independence phase (particularly during the period of 1950 - 1990) was enormous and restoration of it is not an easy task.

Madhya Pradesh is the home of over 25.00 per cent of the nation's tribal population and about 40.00 per cent of the total geographical area of the state is covered by the tribal sub-plan with a bulk of this area being scheduled area, the dependence and pressure on forests are necessarily heavy and enormous. Till 1951, there were four types of forests in M.P. - (a) reserve forests, (b) protected forests, (c) malgujari forests and (d) small teak forests under the Revenue Dept. There were also the forests on private lands, recorded as *malik-makbuza timber.* The *nistar* needs of the people were met from the protected forests and the malguzari forests. In 1951, "when proprietary rights were abolished, the malguzari forests vests in the state. The legal status of these forests were altered, but neither the Revenue Dept nor the Forest Dept. acted to preserve the resource... between 1951, when property rights were abolished, and 1961, when the forests were transferred to the Forest Dept. almost all forests were destroyed" (Ramanathan, 2005:2).

The disappearance of the resource in the village commons (i.e. common property resources meant usually for common use and community purpose, accessed freely by the villagers) has exerted the pressure on the reserve forests, resulting in the practice of illicit felling of trees, grazing of cattle, collection of minor forest products, etc, done by the people are a common phenomenon within the boundary of sanctuaries/biosphere reserves. Thus, common property resources (CPRs) are an integral aspect of the social and institutional arrangements made to meet the everyday requirements of the village communities, specially relevant to the landless, the agricultural labourers and the rural artisans. In M.P. where irrigated land is a small proportion of the cultivated area and lands are often left fallow after a single crop, the dependence on the common property resources to fulfil subsistence needs are pronounced.

Destructions caused to forests by man in this part of the country have created real problem in its restoration work as ever increasing population pressure culminated with lack of alternative economic activities other than agriculture, landlessness, shortage of agricultural land for sustenance, unawareness about importance of forests, etc., have jeopardised the situation. Moreover, the agrarian economy of the area has become static since last few decades. So it is obvious that the people living inside and outside the forest areas have no alternative avenues to earn their subsistence other than to exploit available forest resources.

With damage of forests over a period of last two-three centuries the perennial character of watersheds spread in the area is bound to be depleted to some extent. For instance, in recent past a perennial stream of the Pachmarhi town, locally known as *Bainganga Nala* has lost its virginity, rather dried as reported by many residents of Pachmarhi. And now this is being used for draining for sewage, garbages etc, of the town. The productive capacity of the PBR has declined considerably due to loss of soil nutrients, humus contents and also loss of regeneration process. Even degradation has reached such level that availability of fuel wood has become scarce now.

Apart from this, grazing activities directly causes soil erosion, regeneration of valuable species like teak, sal etc. The very top soil has loosened up and lower layers are compacted. The loose layer is quickly washed away in the rains that has led to reduce the seepage of water into earth. Thus, the period of moisture availability in soil has come down to great extent that also causes stunted growth of the plants. Increasing cattle population and shrinkage of grazing land in the villages have aggravated the situation. Finding no alternative, the cattle owners take the opportunity to graze their cattle in the forest.

For assessing the actual nature of ecological degradation in the PBR area both afforestation and deforestation ratio, no. of existing and drying watersheds, river system, nature of water percolation into soil, groundwater level, rainfall, area of arable land, human and livestock population etc. needs to be seen from historical perspective i.e. the facts should be examined diachronically.

Water Resources

The PBR is basically a part of the central highlands of our country having undulating terrains, endowed with a good number of major and small natural watersheds found in the form of streams, falls, rivers, swamps and also rich in biodiversity. Within its jurisdiction a large number of villages are found to exist, inhabited by the tribal and non-tribal population. As the physiography of the area is undulating in character major parts of rain water is runoff through slopes to rivers, streams, gorges, swamps, etc. During monsoon the people of the area do not feel much difficulties in managing water needs used for different purposes. After monsoon water requirement gets high particularly when cultivation of rabi crops starts. The available water resources are apparently inadequate to meet the irrigation needs of the farmers as well. In dealing with this aspect rainfall and temperature of the area has to be dealt for having an outline idea about the prevailing water situation actually exists.

Rainfall and Temperature

Rainfall and temperature have direct influence over watershed. As the PBR particularly the hilly part receives adequate rainfall, its substantial percentage goes wasted as runoff due to its undulating character. The incessant rainfall makes the weather cool and also leads to adequate vegetative cover in the area. The topographical condition of the area indicates that a good number of rivers, falls, springs, streams and rivulets are found to originate from the Satpura hill ranges and its plateaus.

The rainfall data of three constituent districts of the PBR envisage that no sharp variation exists as far as average rainfall is concerned. While in the hilly part (Pachmarhi hills & plateau, hills of Chhindwara and Betul districts) rainfall is more than the plains

Table 4.6

Average Rainfall during Monsoon (Jun. - Oct.) and Percentage of Precipitation during Jun.- Sept.

Sl. No.	*Name of the district*	*Average rainfall(in mm.) during Jun.-Oct.*	*Average rainfall(in mm.) during Jun. - Sept.*	*Percentage of precipitation during Jun.-Sept.*	*Remarks*
1.	Hoshangabad	1226.94	1208.74	98.52	Based on data available during 1964 - 1976
2.	Chhindwara	966.00	914.40	94.66	Based on data available during 1952 - 1988
3.	Betul	1086.40	1020.30	93.92	Based on data available during 1951 - 1970
	Average	1093.11	1047.81	95.70	

areas. The average annual rainfall of Hoshangabad, Chhindwara and Betul is 1271mm, 1070 mm. and 1172 mm. respectively and in Pachmarhi is around 2000 mm. The rains usually begin in the month of June and continue till the middle of October. But intensity of precipitation is more during June to September and during this period average rate of precipitation is around 90 to 95 per cent (Table 6.6).

It appears from history that in the past rains were uninterrupted in the hills of Hoshangabad district (i.e. Pachmarhi and Bori areas) from the middle of June to August. There was not a single break of any duration (Hoshangabad district gazetteer 1908:19). But in course of time such trend has changed considerably due to various reasons. In this context one point may be relevant to note that no sharp rainfall variation does exist even during the British phase as seen in the past data (Table 4.7).

Table 4.7
Average Annual Rainfall in Three Districts

Sl.No.	*Name of the district*	*Year*	*Rainfall (in mm.)*
		1867 - 1868 to 1905 - 1906	1193.80
1.	Hoshangabad	1927 - 1946	1271.50
		1947 - 1962	1266.60
		1964 - 1976	1271.50
		1951 - 1970	1070.74
2.	Chhindwara	1980 - 1981 to 1990 - 1991	953.04
		1901 - 1950	1130.00
3.	Betul	1952 - 1988	1172.40

Source : Inventory Reports of Forest Resources of M.P., Forest Survey of India, Central Zone, Nagpur; M.P. District Gazetteer-Betul district & EPCO's Project Document on the PBR, 1996.

Apparently average annual rainfall in three districts almost remained constant. But in Pachmarhi and Bori areas of Hoshangabad district which are important part of the PBR, average rainfall has reduced to a great extent during the period of last 50 years (1927-1976) due to various reasons (Table 4.8). Causes of declining trend of rainfall may be : deforestation, urbanisation, grazing, mining and other human activities. Deforestation may have given a major thrust to the declining trend of rainfall in said area.

Table 4.8
Average Annual Rainfall in Two Rain gauge Stations of Hoshangabad District during last 50 years (1927 - 1976)

Sl. No.	*Name of the station*	*Year*	*Rainfall (in mm.)*
1.	Pachmarhi	1927 - 1946	2315.90
		1947 - 1962	2190.00
		1964 - 1976	2050.60
2.	Bori	1927 - 1947	2617.20
		1947 - 1962	2231.80
		1964 - 1976	2070.05

Source: Working Plan of the Hoshangabad Forest Division for the year 1987 – 88 to 2001 – 02.

Like rainfall, temperature, another vital segment of climate of the area that plays significant role in keeping the watersheds alive for a longer period. In other words both rainfall and temperature are interwoven to keep the weather bearable to the whole life system of the earth. The prevailing average temperature in the PBR during hottest months (April to May) and coldest months (December to January) is 36.0° C - 39.0° C and 5.0° C - 9.0°C respectively.

Existing Watersheds/Water Resources

The rivers, natural springs, streams, rivulates etc, originated mainly from the Satpura hill ranges and plateaus (Table 6.9), are the major water resources to the people (mostly Adivasis) of the PBR area. These water resources are considered as a lifeline to them, perceived as community property, and share the same since time immemorial. Such conventional use of these resources leads them to establish their community right, though not under legal framework, entertained by the modern state and polity.

Table 4.9
Major Rivers in Hoshangabad District

Sl.No.	*Name of the river*	*Origin*	*Tributaries*	*Remarks*
1.	Narmada	Summit of the plateau of Amarkantak in Sadol district	(i) Tawa (ii) Ganjal (iii) Dudhi (iv) Chota Tawa (v) Shakkar (vi) Sher (vii) Hiran (viii) Banjar	The Tawa and the Ganjal flow mainly in Hoshangabad, Chhindwara and Betul districts and the rest flow outside the boundary of the PBR (i.e. in Nimar, Mandla, Narshingpur, Jabalpur districts etc).
2.	Tawa	Satpura Range (rises in Chhindwara district and flows down from Betul, enters the Hoshangabad district at the border of Hoshangabad and Sohagpur tahsils and meets the Narmada at Bandarabhan near Hoshangabad)	(i) Denwa (ii) Malni (iii) Gurgugona (iv) Sukhtawa	Denwa and Malni tributaries of the Tawa also fall in Hoshangabad district vis-à-vis in the PBR.
3.	Denwa	Satpura Range (rises in the eastern slope of the Pachmarhi hills that encircles the Mahadeo hills and meets the Tawa at Chormalni)	(i) Sonbhadra (ii) Nag Dewali	Like Tawa, Denwa and its tributaries flow in Hoshangabad district, considered the important natural water resources of the PBR.

(Contd.)

(Contd. Table 4.9)

Sl.No.	Name of the river	Origin	Tributaries	Remarks
4.	Ganjal	Satpura Range (Rajabhorari hills and meets the Narmada at Gondagaon)	(i) Moran	Ganjal flows in the periphery of the PBR.
5.	Dudhi	Satpura Range(rises in the Chhindwara jagirs and flows into the Narmada)	?	Like Ganjal, Dudhi also flows in the periphery of the PBR.
6.	Palakmati	Satpura hills	?	Palakmati, a small river flows in Hoshangabad dist.
7.	Hatara	Satpura hills	?	
8.	Sonbhadra	Satpura Range(rises in Chhindwara, flows down the western slopes of Dhupgarh,a place of the Pachmarhi plateau)	(i) Bori	Sonbhadra and its tributary flow down into the Denwa

N.B.- All the rivers and most of their tributaries flow in Hoshangabd district, play vital role in maintaining ecological balance of the area and various economic pursuits of the people are also influenced by the existing river system.

Of all the rivers and their tributaries listed in the table, Narmada, is the biggest perennial river in whole of central India, it flows westwards and drains huge masses of water coming from the hills, traverses through 1312 kms in the states of Madhya Pradesh, Maharashtra and Gujarat. Many rich and ancient cultures thrived on its banks and been nourished by its waters. Then river, does not fall exactly within the PBR, it flows in the periphery of its boundary and plays most significant role in fulfilling the water needs of the people of the vast areas of central India. The Narmada and its tributaries (Tawa, Ganjal etc), Denwa and other rivers, springs, streams, rivulets etc and the forests as a whole constitute the Great Central Indian Watershed as identified by the Britishers in the middle of the 19th century, that directly influences the ecosystem of the area. The Britishers felt its importance that gave rise to the idea of forest conservation in Central India.

The study highlights that almost every village barring Binora and Chitor Pattan (located in the bank of the river Denwa) is endowed by some natural springs/streams perennially flow nearby their habitations (Table 4.10). The people of all the villages are practically sustained by these natural water resources. The names of the springs/streams listed in Table 4.10 are indigenous type, perceived by the people themselves generation after generation, bear some eco-sense.

If we look into the history of human settlements in different phases of our civilization, water resources are first considered the pre-condition of any settlement in any part of the globe. The study based on core zone villages also testifies this notion.

In the present context people are the centre of study and existence of all other living units or life forms of biodiversity largely depends on the wisdom of the people. The people of the study area are basically the marginalised and disadvantaged section of the Indian society. Their water needs are very limited as far as their living conditions and economy are concerned. The locations of their villages are primarily within the very vicinity of the available water resources i.e. nearby the rivers, springs or streams as shown in Table 4.10.

The actual status of different water sources (Table 4.11) indicates that barring the river (the Denwa) and natural springs/

Table 4.10

Natural Water Resources Available in Study Villages

Sl. No.	Name of the village	Type of natural water resources				Remarks
		River		Spring/Stream		
		No.	Name	No.	Name	
1.	Bariaam	-	-	-	-	-
2.	Dhana Pagara	-	-	1	Sidi Nadi	Perennial type, originates from Mankideo hill located in the southern side of the village. A stop dam built for the supply of drinking water and irrigation purpose; command area-24 acres and beneficiaries-12 cultivators.
3.	Badkachhar	-	-	3	(i) Bhuluk Nadi (ii) Budu Nala (iii) Arapatinala	The first one originates from nearby hill and on it a stop dam built few years back mainly for irrigation purpose, located at north-eastern side of the village; distance-about 1 km., the second one flows in eastern side, gets dry in summer months and the third one originates from south-eastern side near Bilakharra hamlet, also gets dry.

(Contd.)

(Contd. Table 4.10)

Sl. No.	Name of the village	Type of natural water resources				Remarks
		River		Spring/Stream		
		No.	Name	No.	Name	
4.	Kajri	-	-	3	(i) Bade Nadi (ii) Dhunda Nadi (iii) Chikhaljo- wan (Nala	Barring the last one first two are perennial type, considered as lifeline of the village. On Dhunda Nadi a stop dam built recently, has eased the life of the villagers to some extent as far as drinking water and irrigation are concerned.
5.	Rorighat	-	-	2	(i) Chaunahar Jhore/Nala (ii) Kendeaam Jhore/Nala	Both streams are perennial, the first originates from Chunardeo hill about 0.5 km.away from the village and other one flows from the eastern side.Two stop dams built on them for supply of drinking water and irrigation purpose.
6.	Binora	1	Denwa	-	-	The Denwa draws the south and western boundary of the village, used as the principal water source to the people.

(Contd.)

(Contd.Table 4.10)

Sl. No.	*Name of the village*	*Type of natural water resources*				*Remarks*
		River		*Spring/Stream*		
		No.	*Name*	*No.*	*Name*	
7.	Chitor Pattan	1	Denwa	-	-	The Denwa also flows quite proximity to the village and considered as lifeline to the people of Chitor Pattan and its neighbouring villages. Its basin is extensively cultivated for rabi crops.
8.	Ghodanar	-	-	3	(i) Budhgayaam Nala (ii) Khatiaam Nala (iii) Khubrijowan Nala	Three are perennial, originate from the hills located at the southern side of Ghodanar. The first two are used for the purpose of drinking water supply and irrigation. A stop dam exists in the Khatiaam Nala. The third one still remains untaped.

Table 4.11
Status of Different Water Sources in Study Villages

Sl. No.	Name of the village	Type of water source						Status/Remarks
		Tank	Well	Hand pump	River*	Spring/ Stream*	Total	
1.	Bariaam	1	3	1	-	-	5	Out of three wells two are owned by Mrs. Arundhati Roy, the famous writer and social activist for her personal use. Other one is community type formed naturally used by the villagers. The handpump is out of order. The tank is also formed naturally meant for bathing, washing and cleaning of utensils and also for the livestock.
2.	Dhana — Pagara	-		1		1	2	The handpump is out of order.
3.	Badkachhar	-	5	3	-	3	11	One community well, dug in 1937, meant for drinking purpose only (now almost abandoned) and the rest four are owned by four persons meant for irrigating their own land. All hand pumps except one are out of order.

(Contd.)

(Contd. Table 4.11)

Sl. No.	Name of the village	Type of water source						Status/Remarks
		Tank	Well	Hand pump	River*	Spring/ Stream*	Total	
4.	Kajri	-	1	1	-	3	5	Well and hand pump are meant for drinking purpose (almost abandoned now).
5.	Rorighat	-	1	1	-	2	4	The well excavated in 1936(?) is abandoned now. The handpump is also out of order.
6.	Binora	-	-	1	1	-	2	The handpump is abandoned as its water is unfit for drinking due to excessive iron content.
7.	Chitor Pattan	1	-	4	1	-	6	Tank is for agriculture only, but gets dry after monsoon. Of all the handpumps, three are still in operation, used for drinking and other domestic purposes only.
8.	Ghodanar	-	-	1	-	3	4	The handpump is namesake only as its water is unfit for drinking.
	Total	2	10	13	2	12	39	

* Name & status of river & springs/streams have been mentioned in Table 4.10.

streams all other sources (tanks, wells and handpumps), on an average, are of no use due to lack of proper maintenance and management. The handpumps, found in majority cases, are either out of order after few months of their installation or abandoned due to maximum iron content and bad taste. The community wells of Badkachhar, Rorighat (dug during the British period) and Kajri (dug after independence), are full of filthy substances and water of these wells is now practically unfit for human consumption. In fact, the existing rivers and the perennial springs/streams are still considered most precious and principal water sources to the people of all the core zone villages. Sustainable use of all water resources depends on wisdom and knowledge of the people concernd and their method of conservation may also be highlighted.

Conservation and Management of Watersheds/Water Resources

Conservation and management of natural water resources, as today, are much debated among the policy makers, researchers, environmentalists, social and political activists. The watershed management gets more importance now, though not new to the rural communities. The people have their own perception how to tap and conserve natural water resources for sustainable use. But their wisdom and knowledge as perceived and nurtured by them centuries after centuries has disappeared due to influence of modern economic forces and lack of state patronage. Traditional watershed management is still conspicuously noticed in many tribal and non-tribal communities living in forests, arid and semi-arid locations of our country. The devices they use for taping/harvesting rain or groundwater for drinking and agricultural practices, are quite eco-centred and not responsible for damaging the carrying capacity of our mother earth, are marvel to the practitioners of modern science and technologies.

In modern sense, watershed management implies the proper use of land and water resource of a watershed for optimum production with minimum hazard to natural resources. In fact, watershed management primarily leads to resource generation and resource utilisation. The main objectives of watershed management are :

- To protect, conserve and improve the land resources for efficient and sustained production,

- To protect and enhance groundwater storage, increase irrigation and conserving rainwater for cultivation that mitigate drought situation,

- To control and utilise runoff for usual purposes,

- To utilise the natural local resources for improving agriculture and allied occupation so as to improve socio-economic condition of the local people.

The objectives of watershed management mentioned above are almost similar to the notion of the people on whom the present study is based, though not codified by them in the sense of scientific framework. But they are not differed from its basic concept. Truly speaking their wisdom and knowledge are very much within the realm of ethno-science that are significant in the sphere of management of natural resources. Practically the acumen they possess are eco-culturally based, meant for sustainable use of all kinds of natural resources available to their immediate locale or environment not only for themselves but also for the posterity.

The people of the study villages, by and large, try to conserve the watersheds mainly by protecting the tree groves (which are economically useful to their sustenance), available mainly in the catchment areas of different watersheds as seen in Bariaam, Dhana Pagara, Badkachhar, Rorighat and Kajri (Table 4.12). While in the case of Binora and Chitor Pattan people conventionally use the river water for their livelihood.

The rivers and other natural water resources are nature gifted. Hence the people living nearby the river and other watershed areas perceive them (watershed) as their community property, used in sustainable manner since ancient time. The people on whom the study is based are in the lap of nature and their intervention on water and other natural resources available in their habitats are minimal and sustainable in kind. Hence by tradition they never pollute or damage these resources. On the contrary they treat these resources as most sacred and in majority cases the seat of their

Table 4.12
Conservation and Management of Watersheds/Water Resources in Study Villages

Sl. No.	Name of the Village	Method/Nature of Conservation	Management
1.	Bariaam	Nearby the community well (formed naturally) of the village, a small watershed area exists from where water percolates to the said well that never gets dry. This watershed remains alive throughout the year due to the existence of a tree cover formed by several tree species viz. *aam, jamun, umar, khinni, khair/kher, tendu, mahua* etc, are never cut by the villagers.The tree cover is one of the essential components of any watershed that retards the evaporation rate, useful to retain moisture content of the soil. The people perceive the water as god gifted and vital for life and god lives near the water source. This perception also leads them to make the abode of their village deity, familiar as *'Siddhababa'* near by the said well. The place is treated most sacred and the standing trees of the aforenamed species found in the watershed area are protected and conserved by the villagers for their existence.	Community level, culturally bound and eco-centred in character on which their survival largely depends.
2.	Dhana Pagara	A watershed exists in the hills, about 0.5 km. away from the village and water percolates through some invisible springs that passes into small drains and get the shape of a stream, locally knowns as *'Sidi Nadi'* (see table 4.10) on which a stop dam exists. About four decades back three persons of the village took initiative to tap water of the Sidi Nadi for irrigating their field by making kachcha *naha*r (channel), locally known as *dna*r from Panghat (where the stop dam	Community level, eco-oriented and culturally bound that establishes their community right.

(Contd.)

(Contd. Table 4.12)

Sl. No.	Name of the Village	Method/Nature of Conservation	Management
		exists now) to the village. After construction of stop dam the same *nahar* is being used as diversion channel, lies almost parallelly in hill slope, through which water reaches directly to the agricultural field. The channel is built with stone slabs and *morum* (laterite) soil, damaged sometimes during monsoon period, gets repaired by the villagers themselves almost every year after monsoon. To keep the stream alive perennially, the tree groves consisted of *aam, jamun, khinni, mohua, goring, bargad* and other varieties of trees available nearby the stop dam and its watershed area are protected and never felled down by the people. Near the stop dam a *pabitra sthal* (sacred place), locally familiar as *Panghat*, where a stone slab placed under a baniyan tree is ascribed and worshipped as their village deity, known as the 'Panghat Devi' and the aforenamed groves nearby the place are also treated as sacred.	
3.	Badkachhar	Three streams, locally familiar as '*Bhuluk Nadi*', '*Budu Nala*' and '*Arapatinala*' are found in the village of which the first one is perennial and a stop dam, popularly known as 'Marghatai Dam', built on it by the Govt. for the purpose of irrigation. Its watershed is encircled by the forest and the water percolated from it passes through one main drain that gets the shape of a spring/stream after some distance. Its water is lifted by the electric generator and supplied to the village field through PVC pipes.	Also community level and eco-cultural based

(Contd.)

(Contd. Table 4.12)

Sl. No.	Name of the Village	Method/Nature of Conservation	Management
		The water line covers a distance of about 1 km. from the dam site to the village field. Only a couple of families are the beneficiaries of this facilities as far as irrigation is concerned. Because water available from this stream is not adequate to cover the entire crop land of the village. The tree species such as *aam, jamun, khinni, bargad* etc found in the watershed area including the stop dam site are conserved and protected by the villagers. No formal village committee functions to look after the maintenance of pipe line and generator. Operation of generator rests in the hands of the villagers. However, the Panchayat member and the village elders play a key role in decision making for sharing water for cultivation of wheat crop that needs three times watering from sowing to the period of maturity. Water lifted by electricity operated pumpset is absolutely tax free. The Budu Nala after monsoon gets dry and its watershed area is almost devoid of forest cover. While on the Arapatinala a small check dam or *bandhia* exists, built by one person of the village for irrigating his own land about three decades back and managed by himself for its maintenance. Near the bandhia some bushes and a few trees are available which are also conserved by himself. Prior to existence of dug wells, handpumps and stop dam the people of the village used to meet their water demands from the aforementioned springs.	

(Contd.)

(Contd. Table 4.12)

Sl. No.	*Name of the Village*	*Method/Nature of Conservation*	*Management*
4.	Kajri	Three streams viz. '*Bade Nadi*', '*Dhunda Nadi*' & '*Chikhaljowan Nala*', flowing proximity to the village, are the principal water resources to the people of Kajri. The catchment areas of watersheds, covered by *sal, sagoon* and other species, are in hill slopes, located at some distance (about 2-3 km.) from the village, also conserved and protected by the people. In the past the Bade Nadi, flows quite lose to Kajri, it was used mainly for the purpose of drinking water, bathing and domestic uses, but not for irrigation as their agricultural practices were exclusively dependant on rainfed condition. In due course when practice of rabi cultivation (mainly wheat crop) was introduced in the village lift irrigation started. For this purpose a diesel pumpset was supplied to the villagers by the Govt. But this had the limitation to irrigate the whole agricultural land of the village. Now water lifting from stream is abandoned. Water scarcity for agricultural activities still prevails in the village though one stop dam came into existence recently on Dhunda Nadi (see table 4.10). A pipe line has been laid from stop dam to the village and water is stored in a large pucca tank made in the hillock from where water flows automatically to the field through pipes. Its command area for irrigation is also limited.	Community level and eco-centred.

(Contd.)

(Contd. Table 4.12)

Sl. No.	*Name of the Village*	*Method/Nature of Conservation*	*Management*
5.	Rorighat	To the people of Rorighat two perennial springs - *'Chaunahar Jhore'* and *'Kendeaam Jhore'* are the major water resources. As water is naturally available from these springs no such method or device was adopted by them in the past to conserve water resources. Agricultural practices were absolutely rainfed in nature. So they used to need water only for drinking, cooking and other domestic purposes and also for livestock population. The people themselves report that in the past they used to collect from the down stream of Chaunahar Jhore or Nala, locally familiar as *'Gogaiaam'* which has been abandoned recently due to coming up of a stop dam made near its source located in the *'Churnadeo Pahad'*(see table 4.10). Like other study villages the catchment areas of watersheds from where water percolates to these springs/ streams are also covered by mixed forests (teak, sal & other varieties), also protected by the villagers for their cause. With the construction of two stop dams on the aforenamed springs/streams water is brought to the village through pipeline(laid from hill to down slope) that gets accumulated in some pucca storage tanks from where water is supplied through pipeline to the crop field. Water coming from stop dams are now beneficial to the people as it helps to mitigate water scarcity in the sphere of agriculture and drinking water to some extent.	Community level, tradition based and eco-culturally oriented as perceived by the people generation after generation.

(Contd.)

(Contd. Table 4.12)

Sl. No.	*Name of the Village*	*Method/Nature of Conservation*	*Management*
6.	Binora	The river 'Denwa' is the principal water source to the people of Binora that draws east-west border of the village and passes towards north. This river is endowed by a good number of watersheds located in hills, covered by forests. Along the banks of Denwa a large number of economically useful trees viz. *aam, jamun, mohua*, and other varieties are naturally grown, conserved and protected by the people for their survival. The topographical condition of the village (located in hilly terrain) does not permit the entire village population to use the river water for irrigation. Only two persons, who own electric motor for lifting water from Denwa, are able to cultivate their land (the available agricultural lands are also primarily owned by them as reported by the villagers). To bring water from river side to their field a hanging rubber pipeline, supported by some wooden posts, has been laid by them. Lifting of Denwa water by pump for irrigating hill slope is quite unaffordable and also not feasible to the villagers who are, in real sense, poorest of the poor. Only Govt. can intervene on such issue.	Not specifically defined; but eco-culturally perceived by the people as their common property for its sustainable use.
7.	Chitor Pattan	The river 'Denwa' also passes quite adjacent to Chitor Pattan that drains a huge mass of rainwater coming from the Satpura hill ranges to the river 'Tawa'. This river (Denwa) is practically considered the lifeline to the people of Pattan and other	Conservation and management of its water are not in the hands of the forest community alone. As it is

(Contd.)

(Contd. Table 4.12)

Sl. No.	*Name of the Village*	*Method/Nature of Conservation*	*Management*
		neighbouring villages viz. Kukra, Shera, Tekapar, Sharangpur etc. During monsoon it gets the monstrous shape. It is navigable to some extent in its down stream area. During dry months its water recedes near to its bed. The shelf (or upper bed) is used for cultivation of rabi crops particularly wheat and pulses (gram, moong) by the people of Pattan and other villages. As its soil is clay mixed loamy and alluvial type, no fertilizer is used for growing for crops. Its water is lifted by pumpsets for irrigating the rabi field, located in the river bed. In Pattan five persons own pumpsets, usually given on hire basis at a rate of Rs.250.00 per day (6.00 AM to 6.00 PM). Apart from this, recently one pump house has been built in the village by the Govt. in order to lift water from Denwa for irrigating the whole arable land of the village. This is yet to be commissioned.	most precious natural resource they conceive it as their common property, conventionally used for their sustenance since time immemorial. In other sense it is tradition based and cognitive oriented that are governed and influenced by nature on which their entity largely depends.
8.	Ghodanar	Out flow of groundwater coming from the hills gets the shape of spring/stream as found in Ghodanar. The village is blessed with three perennial springs/streams viz. '*Budhgayaam Nala*', '*Khatiaam Nala*' and '*Khubrijowan Nala*' (see table 4.10*)*. The catchment areas of watersheds from where water percolates and flows into these springs, are located not far away from the village and covered by forests.	Community level, eco-centred and tradition based.

(Contd.)

(Contd. Table 4.12)

Sl. No.	Name of the Village	Method/Nature of Conservation	Management
		Nearby the source of water coming out from the hole or flows through the natural drain mango groves are found, which are protected and conserved collectively, that are not only helpful in retaining the said watersheds but also for their economic use(mangoes are abundantly available for both consumption and selling). Apart from mango, other economically useful trees namely *mohua, ber, jamun, tendu, khinni* etc. are available in the periphery of these watersheds. Like Bariaam and Dhana Pagara, the people of Ghodanar worship the nature God, locally known as '*Siddha Baba*' or '*Siddha Maharaj*' whose abode or seat found under a mango tree is located near the Budhgayaam Nala from where the people of the village collect their drinking water. The place is treated as most sacred and no one is allowed to throw any litter in water coming out from a hole.	

village deities viz. Siddha Baba, Panghat Devi is located either in the stream/river side or in the forest that indicates their dependency on nature. Moreover, their socio-cultural base or structure is largely governed and influenced by natural forces. Thus, conservation and management of water resources are purely tradition based and eco-centred that need to be respected, documented for the cause of posterity as well.

Water requirement:

A standardised pattern of average water requirement in respect of human and cattle population and for cultivation of different crops formulated by the experts, needs to be understood (Tables 4.13 & 4.14). Whether the existing water availability from different sources in the study area is sufficient or not that aspect may be assessed with the pattern of consumption of different ethnic groups and their practice of crop cultivation. But this aspect may be left to the experts who could measure the actual water situation prevailing in the area.

Table 4.13
Water Requirement Human & Bovine Population (Quantity in litres per day)

*Human population**		*Bovine (livestock) population*	*Wildlife population***
Rural	**Urban**		
40	135	30-60	?

* Varies from state to state and even region to region.

** Not known

Table 4.14
Water Requirement of Different Crops as per the Base Period (in mm.)

Crop varieties	*Range of total delta (i.e. depth of water requirement as per the base period)*	*Crop varieties*	*Range of total delta (i.e. depth of water requirement as per the base period)*
Rice	900 - 2500	Tobacco	400 - 600
Wheat	450 - 650	Tomato	600 - 800
Sorghum	450 - 650	Potato	500 - 700
Maize	500 - 800	Pea	350 - 500
Millet	450 - 650	Onion	550 - 500
Bajra	450 - 650	Bean	300 - 500
Sugarcane	1500 - 2500	Cabbage	380 - 500
Sugarbeet	550 - 750	Banana	1200 - 2200
Groundnut	500 - 700	Citrus	900 - 1200
Cotton	700 - 1300	Grape	500 - 1200
Soyabean	450 - 700	Pineapple	700 - 1000

Source : Reddy & Reddy 1999. Efficient Use of Irrigation Water.

Water requirement for cultivation of different crops depends on various factors namely soil types, topographical and climatic conditions, nature of farming methods, type of crops cultivated, etc. However, actual quantity of water for various crops is calculated on the basis of delta and base period of crops (Table 4.15).

Table 4.15
Average Base Period* of Different Crops

Crop varieties	*Base period (i.e. No. of days)*	*Crop varieties*	*Base period (i.e. No. of days)*
Rice	120 - 150	Tomato	70 - 75
Wheat	110 - 130	Potato	120 - 130
Sorghum	120 - 130	Pea	80 - 100
Maize	80 - 90	Onion	90 - 120
Millet	120 - 130	Bean	90 - 20
Bajra	110 - 120	Gram	70 - 80
Sugarcane	170 - 180	Cabbage	70 - 80
Sugarbeet	70 - 80	Banana	150 - 200
Groundnut	100 - 110	Citrus	120 - 150
Cotton	130 - 150	Grape	150 - 200
Soyabean	90 - 110	Pineapple	150 - 200
Tobacco	120 - 150		

* Base period of different crops depends on soil fertility, types and climatic conditions of an area or a region.

Delta is nothing but the depth of water provided to the crop from the date of sowing to the date of harvest. And the delta is based on the variety of crops and its base period and it amounts to increase or decrease of number of watering. It is also noted that watering is subject to soil condition of the crop where it is grown. The water requirement for crop is = the total delta x area irrigated.

In the study area the available water is mainly used for drinking and cultivation of Rabi crops particularly wheat that needs average three waterings from the date of sowing and the date of harvest. It is noticed that apparently water scarcity prevails in all study villages as far as cultivation of rabi crops is concerned. Because sizeable agricultuaral lands in each study village barring Bariaam (where the practice of agriculture does not exist) remain vacant due to dearth of water as observed during field investigation,

Table 4.16
Land under wheat cultivation in Study Villages

Sl. No.	Name of the village	Total village area (in acre)*	Total agricultural land (in acre)*	Per cent	Area under wheat cultivation (in acre)#	Per cent
1	Bariaam	12574.06	54.71	0.44	-	-
2	Dhana Pagara**	?	?	?	4.50	?
3	Badkachhar	4582.36	93.04	2.03	8.00	8.60
4	Kajri	2485.26	3.06	0.12	26.25@	?
5	Rorighat	2049.98	19.89	0.97	14.75	74.16
6	Binora	1888.38	45.12	2.39	1.50	3.32
7	Chitor pattan	568.34	79.39	13.97	?+	?
8	Ghodanar	32.00	14.41	45.03	4.50	31.23
	Total	**24180.38**	**309.62**	**1.28**	**113.00**	**19.22**

** Dhana Pagara falls under the core zone of the PBR, (but the same is a prominent part of Pagara village of buffer zone), bearing the features of an ideal village covered for the purpose of the present study.

* Figures are based on 1991 census; # Figures mentioned in column as per the present study; @ Some amount of culturable wastes might have been brought under wheat cultivation due to irrigation facilities in recent time; + Wheat cultivation does not exist in the available cultivable land of the village, but an area of 53.50 acres of the Denwa river bed has been brought under rabi cultivation (mainly wheat crop).

even that also reported by the people. Unlike the past the villagers are now interested to introduce other crops along with wheat cultivation as reflected in Table 4.15. Crops like gram, pea, mustard, moong (green gram), etc, are being introduced on experimental basis seen in different villages. Even the traditional wheat variety has been replaced by the hyv varieties. But this process of crop cultivation is being hampered due to scarcity of water during rabi season. Of all rabi crops wheat gets much more importance than any other crops as it is considered now as staple crop of the people. But the dearth of adequate irrigation facilities hinder them to bring major part of available land under wheat cultivation (Table 4.16).

The table indicates that out of total agricultural land available in all the study villages only around 20.00 per cent of it has been brought under wheat cultivation. This reflects the prevailing water situation experienced by the people of the area. The people also opine that they want to augment their family income by cultivating cereals, vegetables and other cash crops if assured irrigation water is available to them after monsoon. But their hope has not been materialised due to aforenoted reason.

Extension of irrigation infrastructure is considered to be an important factor in changing agricultural practices. In many PBR villages and elsewhere in Madhya Pradesh the tribal cultivators have been supported by the Jeevan Dhara scheme through which the farmers have got financial assistance for digging of wells. Moreover, under the Rajiv Gandhi National Watershed Mission construction of stop dams, ponds etc, have been constructed in many villages of forest areas. The stop dams constructed in various study villages do not serve purposes of the entire population/ whole farming families. It has been reported by many that relatively well-off/influential families (who have interactions with the Govt., NGO & Panchayat officials and members) enjoy the fruits of irrigation facilities to some extent. They also reported that in building the irrigation infrastructures (stop dams, wells, ponds) there was no genuine attempt to involve people's participation and the structures built in different villages did not bring the benefits to the whole village population. The study indicates that cultivation of wheat and other rabi crops are primarily confined to limited families of

Table 4.17

Distribution of Cultivating Households/Families (Rabi crops) in Study villages

Sl. No.	*Name of the village*	*Crop wise distribution of Cultivating households/families*									
		Wheat		*Mustard*		*Pea*		*Gram*		*Moong*	
		No.	*%*	*No.*	*%*	*No.*	*%*	*No.*	*%*	*No.*	
1.	Bariaam	-	-	-	-	-	-	-	-	-	-
2.	Dhana Pagara	10	43.48	3	13.04	4	17.39	-			-
3.	Badkachhar	6	16.67	-		-		-			-
4.	Kajri	21	70.00	-		-		2	6.67		-
5.	Rorigahat	24	54.55	2	4.55	-		1	2.27		-
6.	Binora	3	7.32	-		-		-			-
7.	Chitor Pattan	41	85.42	10	20.83	-		42	87.50	1	2.08
8.	Ghodanar	17	100.00	4	23.53	3	17.65	-			-
	Total	**122**	**46.92**	**19**	**7.31**	**7**	**2.69**	**45**	**17.31**	**1**	**0.38**

N. B.- Crops like mustard, pea, gram, and moong are being cultivated along with wheat by some cultivators on experimental basis, but only for the purpose of family consumption and the yields are quite meagre in quantity.

each village because of paucity of required infrastructural facilities for irrigation, and for societal and other factors (Table 4.17).

Of the total households/families over 50.00 per cent families did not practise rabi cultivation (wheat crop) as reflected in the Table 4.17. While the people of Chitor Pattan practically harness the Denwa water for rabi cultivation practised in the Denwa bed. Water is lifted by hired pumpsets and the expenditures are borne by the individual cultivators who sow the crops.

Under this circumstances an undercurrent of tension and conflict, though not overtly flared up due to kinship bond or for other reasons, has brewed among the fellow villagers. No ventures are yet to be found to bring the deprived families from irrigation facilities. Communities living in the study area are not so empowered that they can spearhead any mass movement for redressal of their cause or motivated by any activist groups who could guide them for fulfilling their water demand and other grievances. Water scarcity, imposition of various restrictions in harnessing various forest resources under the biosphere rules have forced them to lead miserable conditions in respect of healthcare, education, family welfare and economic activities.

It is conventionally believed that barring agriculture (other than rainfed crop varieties) water consumption among the village community is comparatively less than the urban population. The average requirement of water for an urban person is 135 litres per day though it varies from state to state or area to area, while the average per capita consumption in the rural area is only 40 litres (see table 4.13). In spite of such low consumption rate the rural people are deprived of their minimum requirement for their sustenance. Why this discrimination continues? What is the role of the state in mitigating such abrupt situation as far as per capita water consumption is concerned in the rural sector? These questions need to be debated at this juncture in right forum.

In development paradigm water is considered one of the vital components of human progress and its demand in various sectors is increasing day by day and the PBR area is no exception to this process. As the PBR being considered both religious and eco-tourism destination (where thousands of pilgrims and tourists of other nature throng to visit Pachmarhi and various religious places during *Shivratri, Nagpanchami* celebrations and other seasons) demand for water (by the settled & floating population) has undoubtedly increased considerably in recent time and in future when the tourism schemes (which are under active consideration of the State Govt.) will be promoted in and around Pachmarhi and other places of the PBR the water situation will be more aggravated that may lead to create water unrest among the people of the area. In this regard one example may be cited how the people used to face water scarcity in the recent past. Nagduara, a famous religious spot located in Kajri, a study village, is visited by the people of different states on the occasion of the Nagpanchami and on this occasion a week-long fair is also organised. Visitors and the local people used to face drinking water problem that has been minimised to some extent a few years back with the construction of a stop dam on one of the perennial streams located nearby the village. Despite such situation tourists' inflow in Pachmarhi area increases almost every year (Table 4.18). But infrastructural facilities are not adequate to cope up with the growing needs of the tourists.

Table 4.18
Tourists' inflow in Pachmarhi Area of the Pachmarhi Biosphere Reserve

Year	*Domestic*	*Foreign*	*Total*
1990-91	42216	310	42526
1991-92	144446	310	144756
1992-93	30165	94	30259
1994-95	165891	124	166017
1995-96	368341	80	368421

Source : Madhya Pradesh Tourism Development Corporation (MPTDC), Bhopal.

As Pachmarhi falls under buffer zone of the PBR human interventions in different spheres are increasing day by day that may inevitably affect the ecosystem of the area. Even by promoting eco-tourism environmental degradation cannot be stopped. The rivers specially Denwa, streams, falls etc, are the important water source to the people of the indigenous origin and settled population of the Pachmarhi town. With the tourist flow and other development activities (recently some development activities have come to a standstill due to some public interest litigation cases) water sources are becoming polluted. The entire sewage of the town goes to the river Denwa through nala. The garbages are dumped in the open field located about 2-3 kms away from the town which pollute water sources (rivers, streams and wells), land and forests that affect both human population and wildlife. The town authorities do not have immediate solution of both garbage and sewage treatment.

It is also known that the people of both core and buffer areas (a good number of villages surrounded by the forests in the buffer zone) would be shifted elsewhere as per the biosphere guidelines. But the issue of proper rehabilitation of the affected people is yet to be finalised. And the fate of the hapless Adivasi people hangs in balance. In this context one press report may be cited here about the rehabilitation condition of the displaced poor tribal families of Madhya Pradesh.

"About 100 adivasi families of Ghogrikheda and Dogjharna villages in Hoshangabad district were recently displaced from their land by the forest department to make way for adivasi families evacuated from the Bori Wildlife sanctuary. There were protests, but the displaced folk could do little without ownership rights. Such families are spread over half-a-dozen districts" (Bhan, R. 2005: IV).

From the above cited example it is understood that the affected people are always ill-treated and deprived of minimum needs for their sustenance, remain poor generation after generation. Displacement and rehabilitation issues are now very debatable and complex in character and in the midst of this controversy the affected people mostly the tribals move from place to place for sustenance. And thus conservation of biodiversity, wildlife in the national parks, sanctuaries and biosphere reserves is meaningless to them unless they are properly rehabilitated with adequate infrastructures that may help them to sustain their livelihood.

Migration of tribals and other weaker sections from their ancestral land in search of livelihood is also partly related to water scarcity. Drought and water scarcity situations often compel enmasse migration as seen in different parts of the country. The study area is endowed with a number of natural water resources. But lack of adequate infrastructure for taping, management and utilisation compels the people to face water scarcity visible in the sphere of crop cultivation and other economic activities that also lead them to follow seasonal migration particularly during the rabi season to some other neighbouring areas for employment. They are engaged usually by the affluent farmers (mostly non-tribal landowners) of the valley area of Hoshangabad district and other places at a lower wage rate. Finding no other alternative avenues for employment the tribal workers are compelled to engage themselves at the lower wage for sheer subsistence. This scenario almost prevails uniformly in entire central highlands of India. At this juncture various watershed programmes as launched in various rain-fed regions of M.P and other states need to be implemented also in the study area with the involvement of the local people so long they will be allowed to remain in their locales by the state.

Summary and Findings

In the foregoing pages an endeavour has been made to focus a broad outline about watershed which prevailed in core zone villages of the Pachmarhi Biosphere Reserve (PBR). The study deals with three major components of watershed - land, forest and water. The villages covered for study are having several watersheds from where groundwater outflows and drains into a common outlet that gets the shape of spring or stream, river etc (which are mostly perennial type) on which livelihood of the people largely depends.

The people of the villages are mainly Adibasis by ethnicity, living in forests centuries after centuries. By profession they are, by and large, agriculturists and in agrarian category majority of them are the marginal farmers. The lands which they own are cultivated by themselves and yields which they produce are inadequate to sustain their families. In reality they are still under subsistence level. The forest resources, which were freely available to them perceived conventionally as their common property since ancient time, have now been scarce and their community right on the forests and other natural resources have also been curtailed by the state under the pretext of biosphere reserve. With the declaration of the biosphere their life has become miserable due to various restrictions imposed by the statutory authorities.

As water is the part of a larger ecological system on which the entire life system of the earth depends for sustenance its conservation and judicious use rests upon the man's wisdom and knowledge. But the pattern of global consumerism and so-called development trend leads to make this most precious thing more scarce for which larger poorer, unprivileged and marginalised masses of India and other Third World countries are the worst sufferers for maintaining their livelihood. Dearth of water has become acute in the semi-arid and various highland regions of India that has led to emerge intra and inter-community as well as inter-state tensions and conflicts. In the recent time a slogan on the issue of owning and sharing of water has cropped up among the deprived masses of the country. The slogan is:

"Water for life, water for right, water for people"

But how far this slogan would mould the state policy of development as well as conscience of the people belonging to the higher strata of the society whose consumption pattern of resources is infinite, that puts a big question. In this context, it may also be mentioned that almost all existing water resources available in the country side, biosphere reserves, hills and mountains etc (as conserved & protected directly or indirectly by the forest dwellers and other rural masses from depletion, pollution, mass destruction of forests etc) are being tapped primarily for the town and city dwellers, commercial and industrial establishments. They are the food producers, but deprived of their legitimate share and die in starvation due to lack of food and water.

Unwise material development at the cost of natural resources as west follows, rather, is threat to the very existence of the Nature's whole creations. India's development rests upon the blending of spirituality, wisdom and knowledge and rational use of natural resources as nurtured by the people of India since ancient time, where materialism was meant for the co-existence of man and all other living and non-living things, never created any problem in maintaining the ecological balance of our Mother Earth. But in recent time India's age-old philosophy of development is being altered or dismantled by the state and the people of so-called elite class (business class, industrialists, bureaucrats, intellectuals etc) who exert influence on the country's polity and the apparatuses of governance.

Today's development paradigm is based on the **philosophy of consumerism,** measured in terms of GDP growth, which is practically meaningless to the common masses of the country. The slogan for **Conservation** and **Sustainable Development** is the resultant effect of politics of development. The present development scenario is, in fact, the alteration of God's creation of all living and non-living things that leads to bring God's wrath in the form of various natural phenomena as happened in different parts of the world including India. To save biodiversity and to maintain ecosystem of the earth, the existing philosophy of development needs to be critically examined. Then the usage of conservation of natural resources and sustainable development would be more meaningful, significant and fruitful to the larger common of this great nation.

Findings:

- Watersheds available in the study area are primarily small and perennial type; to maintain perennial character tree groves are found in the source or drainage area from where water outflows are conserved by the people.
- Have the notion to establish community right on available watersheds/ water resources and forest resources on which their livelihood largely depends.
- Management of watersheds rests in community level, culturally bound and eco-oriented strategies that embedded in their socio-cultural base.
- No proper measures are taken to tap adequate water from various watersheds/ water sources that are wasted throughout the year. This leads to face water scarcity as far as crop cultivation is concerned during rabi season. Major part of the arable land remains uncultivated due to lack of irrigation.
- Water scarcity also partly compels the community members to follow seasonal migration particularly during rabi season in the neighbouring areas in search of employment.
- Miserable conditions prevail in the life of the people whose subsistence economy has been shattered specially after declaration of biosphere reserve. As part of biosphere

guidelines they may be shifted from their ancestral land to elsewhere in due course. Displacement and Rehabilitation issues are remain obscured.

- Forests of the PBR have lost its virginity as the same was severely damaged due to rampant felling of trees and grazing took pace for over a period of last two to three centuries.
- Tourism promotion will disturb ecological balance in the sphere of water availability, rainfall, forest vegetation, wildlife, land and soil, agriculture etc.
- Communities living in the study area are unaware about various watershed programmes launched throughout the country by the Govt. as they are not so empowered that they could raise their voice about their cause. No scientific measures are seen from the part of either Govt. or NGOs to protect and conserve the existing watersheds with the involvement of the local inhabitants of the area.

References Consulted

Anonymous 1996 *Pachmarhi Biosphere Reserve-project document.* Bhopal : EPCO.

Anonymous 1997 *Report on inventory of forest resources of Chhindwara district of Madhya Pradesh.* Nagpur : Forest Survey of India, Central Zone.

Anonymous 1997 *Report on inventory of forest resources of Betul district of Madhya Pradesh.* Nagpur : Forest Survey of India, Central Zone.

Anonymous 2001 *Report on inventory of forest resources of Hoshangabad district of Madhya Pradesh.* Nagpur : Forest Survey of India, Central Zone.

Anonymous	?	*Working plan of Hoshangabad Forest Division for the year* 1987-88 To 2001-02. Dept. of Forests, Govt. of Madhya Pradesh.
Bhan, Rohit	2005	Madhya Pradesh-PC, uneasy. *The Indian Express,* Sunday, 8. 5. 2005 : IV.
Buch, M. N.	2005	Killing the inheritors of the earth. *The Indian Express,* 27.5.2005: 9.
Corbett, G. L. and R V. Russel	1908	*Central Provinces District Gazetters-Hoshangabad district,* Vol. A -descriptive.
Forsyth, Captain J.	1919	*The highlands of Central India – notes on their forests, wild tribes, natural history and sports.* London : Chapman and Hall.
Ramanathan, Usha	2005	*Common land and common property resources in M.P.* 2000-04 (working paper). International Environmental Law Research Centre.
Reddy, G. H. and T. Yellamanda Reddy	1999	*Efficient use irrigation water.* Ludhiana : Kalyani Publishers.
Shrivastav, P. N.	1971	*M.P.District Gazetters-Betul district.* Bhopal : District Gazetters, M.P.

5

Ecology and Material Culture of Pachmarhi Biosphere Reserve

Tamal Maity

Pachmarhi Biosphere satisfies the criterion proposed by the UNESCO that each Biosphere Reserve should contain three areas, such as core area, buffer area and transition area. Satpura National park is in the core area and Bori and Pachmarhi sanctuary is in the buffer and transition area respectively. Pachmarhi Biosphere Reserve covers three districts of Madhya Pradesh, namely Hosangabad, Betul and Chhindwara. The total area of Pachmarhi Biosphere Reserve is 4981.00 sq.kms. Out of this area 62.47% is under forest, 29.38% is under cultivation, 4.94% is water bodies and rest 2.18% remains as wasteland. Built up area within this Biosphere Reserve is only 0.54%, which includes urban settlement of Pachmarhi, Babai, Sohagpur, Pipariya, Tamia, etc. The Biosphere Reserve contains 575 villages of which 508 are revenue villages and rest 67 is forest villages. These villages come under ten tahsils of the three above mentioned districts.

Area of Study

The core area of the Pachmarhi Biosphere Reserve is about 524.37 sq.kms. (Satpura National Park) and it is kept absolutely undisturbed. There are eight villages under core area namely, 1) Barium Darmarrha, 2) Dhana Pagara, 3) Ghodanar, 4) Badkachar, 5) Kajri, 6) Rorighat 7) Binora and 8) Pattan. All the above seven villages are under Pipariya tahsil of Pagara Gram Panchayat and Pattan village is under Sonpur Panchayat of Suhagpur tahsil of Hoshangabad district.

Ethnic Composition of Core Area

The core area villages are dominated with the presence of scheduled tribes such as Mawasi and Gonds.

MAWASI

The Mawasi are considered to be a subgroup of the Korku tribe. According to 1981 census the tribal population of Mawasi tribe is 11,093. Mawasi tribe depends for their living mainly on agriculture, forest produce collection, forest labour, agricultural labour, fishing, etc. Their main crops are Jawar, maize (makka), kudo, kutki, etc. The tribe collects minor forest produce like amla, chirongi, mahua, tendu leaves, mahul leaves, honey etc. In the rainy season food grains are not available. They take boiled *mahua* and *kadukanda* as their food.

The Mawasi are an endogamous group. They are divided into number of clans. The clan is known as *got.* The clans are generally totemistic. The boy cannot marry a girl belonging to same clan as they are considered to be descendants of same ancestor.

The Mawasi worship mainly Mahadev of Pachmarhi, but they also worship nature in the form of rivers, mountains, trees, sun, moon and Lord Rama, Krishna, Hanuman, Ganesha, Durga. Laxmi etc. Tradition of "Bali" (oblation) is also found in Mawasi tribe. At the time of pooja they kill /offered goat, hen, and cock and also drink local liquor.

The Mawasi are changing due to the influence of education and communication. There is a significant change appearing in their living standard, dressing, agriculture and other aspects of life. Traditional folk, music, dance and arts are disappearing gradually.

GOND

In Madhya Pradesh, Gond population is 53,49,883 in the 1981 census. They have 20 sub-groups in Madhya Pradesh. Their mother tongue is Gondi, belongs to the Dravidian family of languages. In Panchmarhi, core areas, Gonds speak Hindi language within their family and also outsider. The Gonds strictly follow the norms of endogamy at the group level. The social divisions are based on four Gonds group of phratry commonly known as *saga* which is totemistic in nature and each share a common myth of origin. They have several clans under each phratry. The Gond practice cross-cousin marriage and preference is given to mother's brother's marriage. Bride price (*hunda*) in cash and kind is generally paid to bride's father. Monogamy is the general practice though polygyny is also permitted.

Their traditional occupation is agriculture and animal husbandry have sustained their economy. At present Gonds are settled cultivators, forest labourers, agricultural labourers, wage labourers, etc. Gond women have a role in animal husbandry, collection of fuel and potable water and other economic activities. Their main crops are Kado and Kutki (millet), paddy, jowar, wheat, tuar, urad, gram, etc.

The Gonds have a rich tradition of Art and craft which includes wall painting, body tattooing; etc. They are fond of music and dance and make musical instruments. They participate in Hindu festivals like Diwali, Holi, Dussehra and other religious activities.

Objective of the Study

Present study intended to document the material culture of the core villages of Pachmarhi biosphere reserve area in relation to ecology and ethnic groups of the area to establish the impact of ecology in material culture production as part of man nature interaction complex. In this connection following objectives are fixed.

1. To understand the indigenous technique and usage employed in the manufacture and use of various tools and implements and house building process of the local people
2. Documentation and preparation of a total inventory of the implement used in various sphere of life.
3. To understand local resource used pattern in material culture production.
4. The ethnic affiliation and gender orientation of material culture.

Material Culture

Material culture is the study of those objects (other than human) and all those artifacts with which man surrounds himself in his attempts to ameliorate his position in his environment (Hutton, 1944). Material Culture is the cultural manifestation of material use. It consists of certain process involving visible materials and also the results of such processes. Material culture is visible to the naked eye. It involves the transformation of natural items to a cultural item. The cultural semantics of such items are determined by the ecological context and mode of use ascribed by the community. Every culture allows certain transformation process to define its material use. In general, it is a need based culturising process of the natural items available in the surrounding environment to define the sociology of existence. The transformation process involved in the formation of a cultural item in the respective material culture of the ethnic groups is to be noted as: Natural object >> cultural object by use, Natural items >> object of use by human workmanship, Object of use by possession from outside (non indigenous) and Nativising the possessed object or by maintaining as a borrowed item.

Indigenization of the non- indigenized by adapting the model and recreating it or by using the item exclusively different from the purpose in which it was created by others.

However, the indigenous structure of material culture of any community lies primarily on the first two categories while any item get nativised further by finding new value in the borrowed culture can also be considered as native during the course of time and otherwise it is to be accounted as a borrowed item. An evaluation of the material culture of the above mentioned communities on the said frame reveals that most of the items are natively made and side by side borrowed items are also in vogue.

Importance of Material Culture

(1) Material life is an index of the general level of the knowledge and of the intellectual attainment of the people.

(2) Ecological factors play a great role in shaping the material cultures of the communities.

(3) Man utilizes his natural environment/ natural resources with the technology at his disposal.

Material Culture as part of Indigenous Knowledge

Material culture surveyed in this study is treated as part of indigenous knowledge. In this connection each item manifests its ecological orientation and the social organization of knowledge behind the production and use.

Ecology and Material Culture

Role of ecology in shaping material culture is a domain of cultural Anthropologist's endeavour in understanding the man nature interaction specific to the area and the associated communities. Present study reveals that the ecological surroundings helped the ethnic groups to procure materials and shape them to have an eco-friendly subsistence tool production and an indigenous way of utilizing nature for the betterment of human survival.

Ethnic Structure of Material Culture

In a study of material culture, it is inevitable to make an enquiry that how the material culture is related to a multiethnic village setting as to understand the ethnic boundaries in relation with material culture. Present study observed that as such no ethnic boundary can be marked either in the production or in possession of material culture as far as the area of study is concerned. It is manifested more of a regional characterization beyond the boundaries of villages or ethnic groups as no item is marked as ethnic specific or village specific.

Ethnic Craftsmanship

As far as ethnic craftsmanship of the studied area is concerned, it is not specifically confined to any single group though artisan community like Barai contributes their share of skill sometimes in material culture production. In general, each group is capable enough to organize and execute the works related to production, use and maintenance of the artifacts. Moreover, the presence commonness in material culture across the communities encountered also supports the above observation. Thus, it can be better understood as community free material culture is in vogue as far as studied villages are concerned.

Gender and Material Culture

The material culture reflects gender demarcation in use and not in production. It is observed that even a similar pattern found among the gender bound use among all ethnic groups settled in the core area.

Modes of Documentation

The Material culture is classified and descriptively documented with the help of sketches and photographs in addition to record its measurements and process of production.

The scheme of classification:
The material culture is classified in this study as

(a) Subsitence Related
(b) Domestic Related (architecture, utensils)
(c) Amusement Related (games, recreation, art, music)
(d) Body Related (dress, cosmetics, ornaments)
(e) Religious Oriented (ritual related crafts)
(f) General (Tools in General)

CULTURAL PROFILE AND MATERIAL CULTURE OF CORE AREA –STUDIED VILLAGES

Material life is an index of the general level of the knowledge and of the intellectual attainment of the people. Keeping the above in mind, some of the cultural domains are enlisted under to show a profile of cultural practices and associated material culture prevalent in the core area.

Personal care:

Bathing customs and habits

Most of the people in this area have the habits of taking bath once in a day. Those who are agricultural labour or doing labour work prefer to take bath late in the evening after completing agricultural work. Sometimes they do not have the time to take bath everyday but during festival or rituals they take bath in the canals or tanks and wash their clothes too. Most of the canals are outside their village settlements and they are used for washing clothes and bathing purposes.

Teeth cleaning:

They use the toothbrush, sticks of *nim, babul, jamun*, etc. tree twigs, and powder of cow dung cake for cleaning of the teeth. Those who are economically well off use commercial tooth powder like *Dabur Lal Dante Manjan* etc. Some people are not very particular about this, but they wash their teeth by finger after taking meals.

Toilet practices:

Most of the people go for their natures call outside their settlement a little far away. Women usually complete their daily routine either early in the morning or late in the evening. Sometimes they use grass, leaves etc., for washing after natures call instead of water in the forest.

Spitting habits:

Most of the people have the habit of chewing tobacco, betel and betel nuts and in general they spit outside their house.

Ventilation:

The homes are built with locally available materials and the tribal huts or residential portion is little high above the ground level. For ventilation they prepared one or two windows in their house.

Disposal of garbage:

The garbage is stored in the outskirt of their residence by individual families as they are used manures in the agricultural field. Those who do not have agricultural land, sell the same annually.

Drinking water:

Source of drinking water varies from one place to another. Wells, hand pumps and stream water are the main source of drinking water. During the summer months (April/ May/June) most of the water source gets dry and people are forced to drink the ponds/ canal or river water or fill it for daily use.

Washing clothes

Washing of clothes is a rare practice among the Mawasi and Gond. Some of them even do not at all wash their clothes and change their beddings. This can be depending on the economic condition of them. The labourers wash their clothes once in a week or fortnight whereas those who are economically well clean their clothes time to time round the year.

Diet pattern

Staple food of the people of the core area is *kodo, kutki* (millet), *makka* (Maize) and other supplement foods are pulses like tuar dal and vegetables. They take normal diet like roti of wheat (*gehu*), *makai ki roti*, *jowar ki roti* with rice, *tuar dal*, *methi* leaves, *billori ki vagi* etc., which are available in their kitchen garden and forest. In the morning before going to field they use to take *pej* (gruel liquid food) which is prepared from *kodo/kutki/meg* by grinding in chakki at home and mix with *mahi* (butter milk) and salt. In the evening they consume handmade bread (wheat or corn), vegetables and *tuar* dal. Their occasional food is *mahua* fruits/bijh (*koina*) (*Bassia litifolia*), *chiranji* from achar tree (*Buchanania Latifolia*).The cooking medium of the core area is soyabean oil/ groundnut oil.

Dress

The elder person of the village wear *dhoti* which is worn up to knees, banyan and shirt or *kurta*. The younger generation wears pant and shirt. Womenfolk wear *lugda* or sari and blouse. *Lugda* is a six yard long handloom cloth. No special dress is worn by women during festivals except that new clothes are used.

Ornaments

Ornaments are mostly liked by women. They wear both silver and gold ornaments. Some years back, bell metalled ankle band was prevalent among women which is rarely seen today. In their arms ornament they called *Bakra*, in their neck called *hansli*, in their leg called *tora*, *bichia*. Some male folk wear silver ring in their ear lobe, called *mundari*.

Fire wood collection

During collection of firewood in the forest by men and women, they have some restriction as auspicious tree. Gonds do not break any twig or collect *Saaj ka peir*, Burgat (Banian tree), pippal, Tulsi, Bail/Bill, Dhutra etc. But Mawasi people have two restrictions of Bargat and Pippal as they reported.

They also collect Rambouri grass from forest. After bringing it at home they dry it in their courtyard and prepare broomstick for their own use and sell it in the market @Rs.10/- or Rs.15/- according to size.

House pattern

The houses in a Mawasi settlement are arranged lineally in rows. The houses stand separated a little from one another with a narrow passage between successive ones. In some houses, the front portion is enclosed with a fence made of bamboo wattle (strips). Kitchen gardens are attached to some of the huts. The houses in a settlement may be either two sloped or four sloped, but the latter type occurs more in numbers.

The four sloped huts are bigger in size and height and more strongly build than the two sloped huts. No structure or material differences are observed in the construction of huts for the different ethnic groups living in the same settlement.

The side walls of a hut are usually made by plastering mud over bamboo wattle or slender twigs, which form the core of a wall. Sometimes no mud plaster is put over this structure and bare portion of bamboo wattle or twigs is left as it is. In few cases, composite walls made of both wattle and twigs, as well as patch works of mud over it, are also found. The walls which partition the inner portion of a hut into rooms are also made in the same way.

The thatching of a hut is done with a kind of *mahul* leaves or grasses which grow abundantly in the neighboring forest. Now a day's plastic sheet and tiles are used as thatching materials. The caves of a hut come down slantingly to a height of about 4 to 5ft from the ground. A four sloped hut is, on average, about fourteen feet in height, while in case of a two sloped hut, the height varies from seven to eight feet.

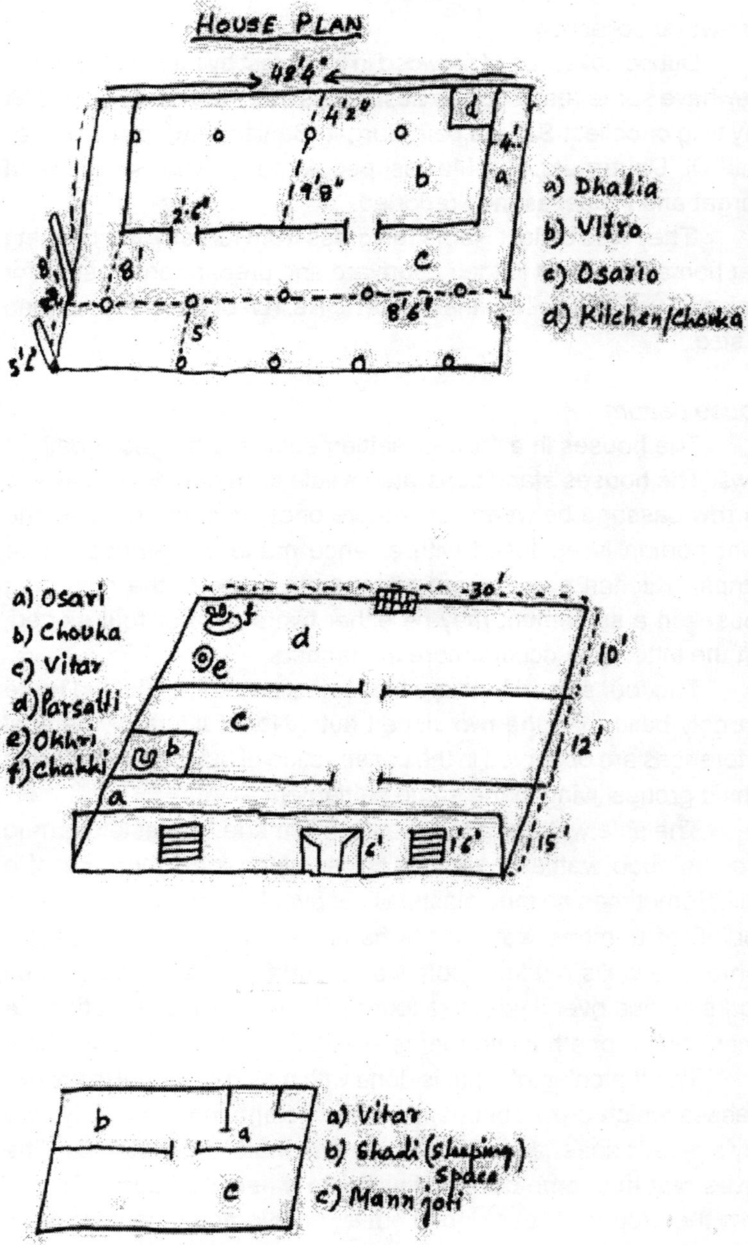
HOUSE PLAN
48'4"
4'2"
d
4'
b
a
19'8"
2'6"
c
8'
8'6"
5'
a) Dhalia
b) Vitro
c) Osario
d) Kitchen/Chouka
a) Osari
b) Chouka
c) Vitar
d) Parsatti
e) Okhri
f) Chakki
30'
10'
12'
15'
1'6"
a) Vitar
b) Shadi (Sleeping) Space
c) Mann joti

Figure : 5.1

The floor is made of earth. The height of the floor from ground level varies from house to house. The floor space inside a four sloped hut is usually divided into one main room for sleeping (*majota*) and kitchen called *vitar*, one store room, and a veranda(*osari*) in its front part, it remains open in front with lateral sides enclosed by the side wall. The entrance to the main room is made through the varanda (*osari*). A few four sloped huts contain two entrances one in the front and other at the back. Osari or verandah is frequently plastered with earth mixed with cow dung paste, thereby giving generally a neat and clean look.

Most of the four sloped huts contain a structure called *manda* which serves as a main storing chamber. It remains at a height of about 7 to 8ft from the floor. Inside a hut, the *vitar* portion especially the kitchen and the store room remain dark as no windows are provided. There is no outlet in the kitchen for smoke to escape the smoke go slowly through the thatched roof.

The utensils are generally cleaned by womenfolk of the house in a corner of the covered courtyard, but most of the households are noted to have erected a platform about three feet in height over which the earthen pitchers or steel containing water of daily use was kept. Some houses are having *mushar* while other contain *mortar* holes (*okhri*).

The grinding stones (*chakki*) are kept on one side of the house. The normal method of storing the food grains is to keep them in a large bamboo grain bin (*palai*) which are plastered with cow dung. This grain bins are arranged in rows on a bamboo or a wooden platform (*manda*) being built on wooden post.

Household Possession

Earthen *chullas* (hearths) are used to cook food in every household. They cooked rice in a steel *ganja* (*handi*) and pulses, vegetables are cooked in ganja or iron made *kadai*. The utensils used for eating purpose are *thali* (dishes) and *lota*. Lota (*charu*) is usually for drinking water. *Dal* or vegetables curry is taken in small plates. Water is stored in earthen pots of black colour. For bringing water from the tank and well, steel vessels (handi) are used.

Visitors are offered cots to sit upon. The cots are made up of wooden frame and stringed with the rope or coir.

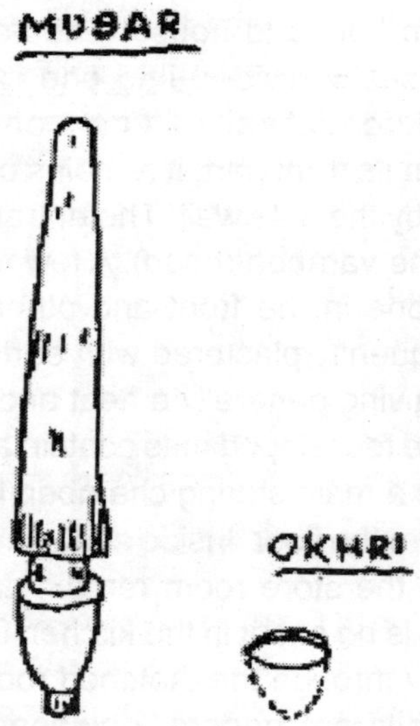

Figure : 5.2

The grain stored in a big earthen vessel known as *kothwa* or *kuthia* is mostly prepared by the men or women. For dehusking *kudo, kutki, siko, saina* they use *musar* and almost every houses has a *musar* and *okri* in their *osari* or covered veranda. It is prepared by the individual itself. The *okri* is bowl shaped pit placed inside the floor. A wooden structure of 43 inches size used for husking known as *musar*. In this pit, *kudo* or *kutki* is kept. By repeated strokes by *musar* husk is separated from the grain. The work is done by womenfolk and is also used to grind turmeric and pound pulses.

Chakki/Chakia

It is a rotary quern found in every house and is usually installed in one corner of the house. It is constructed of two flat stone slabs, one fixed in the earth on which 8 inches iron bar stands which works as an axis to operate the next one. The next slab is fixed with a wooden handle (*muthia*) 6inches height to rotate the slab on the fixed slab. There is a hole (*gala*) in the middle of the next slab through which the grains are put in and pounded by rotating it. This work is done by the womenfolk.

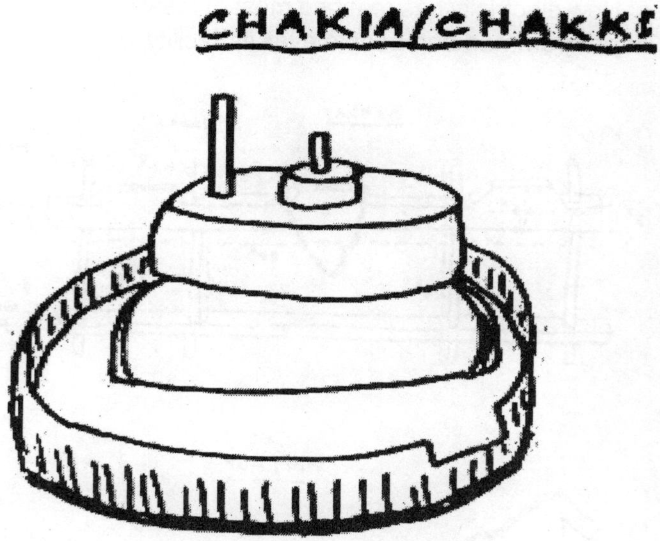

Figure : 5.3

Sil (*chatan*) and *ruria* is another device used to prepare *chatni* and pound chilly, coriander etc. Sil is a rectangular stone slab is 2 ft in length and 1 ft in breadth. The thickness of the sil is usually 3 to 4 inches. The upper part of the slab is made rough by strokes of chisel and hammer. Ruria is a pestle of stone. It is a circular slab of about 6 inches length and 2 inches radius.

Chulla is made of mud in which fire is burnt and the food is cooked. *Chullas* are found to be of two types, *chulla* with one mouth and *ulchulla* with two mouths. It is made of thin plasters of mud. Firewood they collect from forest and *kanda* from cow- dung use for cooking.

Light

The villagers posses' *chimni* in their houses for light. *Chimni* is a small lamp with oil-container and glass. It is used inside the house. *Deeya* is a earthen hollow lamp used during religious worship and Diwali festival. It is filled with oil, and there is a cotton wick, one end of which comes through the lamps lip, while the rest trails in the oil held by the container. In the kitchen they use *deeya* also. Now electricity has reached most of the villages with a single point programmed and people avail its facility.

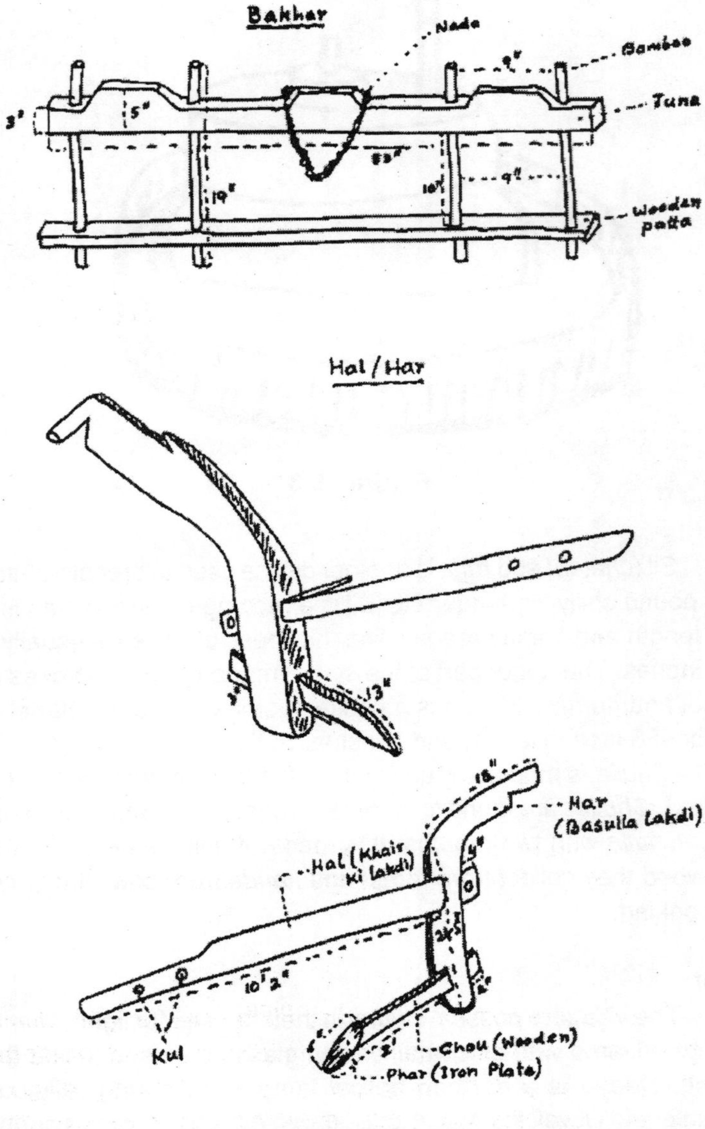

Figure : 5.4

Smoking:

The habit of smoking tobacco is more popular among the villagers whether old or young. Usually tobacco is smoked in the form of *bidi*. Offering *bidi*, is also considered a mark of hospitality.

Horticulture/Harbori Culture:

The plant of mango, banyan, guava peepal, *nim*, *bel*, *jamun* and babul etc. are planted by these people for the economic and religious purposes. Peepal and banyan trees are the abode of gods. Nim and babul are the best tooth brush-stricks. Mango and guava trees provide seasonal fruits. *Bel-patti* is offered to Lod Shiva and its fruits are very delicious.

Agricultural tools and implements:

Land and cattle are the two main resources of the people of core area villages economy. Bullocks have much importance in agriculture.

Tools used in agricultural operation are *har/hal* (plough), *bakhar, patta* (used for leveling the field, *kudari/tangia/dasta* (pick-axe), *dattri/dattre/hasia* (sickle), *gaiti* (for digging), *khoria* (for cleaning mud from *har/hal*), *kolari* (used for separating straws from the grains at the time of threshing) etc.

The common plough is *hal* is an ordinary wooden plough with an iron shave, which varies in weight. The heavier plough certainly produces better crops. The iron shave, called *phar* or *upka,* projects about six inch beyond the wooden boot. To press the share into the ground, the cross-beam of the yoke is loaded with a weight, called *ghangra*, which is made light or heavy according to the depth of furrow required.

A bamboo tube (*por*) is fixed, with a broad wooden cup (*chadi*) at the top, which is fed by a boy or women who follows the plough. Often the hole at the bottom gets clogged with earth, and for a few yards, till the boy finds the tube getting full of seed, the furrow is unfed.

These are the important implements of agriculture, but for his other operations the cultivator is also equipped with tools of recognized shape and name. The *khoria* is the goad, pointed at one end for the benefit of the oxen, and fitte a with a flat piece of iron at the other end to scrape the mud from the plough. The sowing basket is called *supra*. The reaping sickle is known as *dattri*. The implements of threshing are crooked fork, called *ankṛi* or *kolari*, and the wooden rake or *pachanya*, for winnowing the basket (*tokni*), and the broom (*bahari*) are used.

The plough is called Hal. It consists of mainly three parts, a) plough proper b) beam c) yoke.

The plough proper (*hal*) is made of a single piece of wood which is slightly curved at the middle. The handle (*bonta*) is long and is made of the same piece as the body. In the middle of frontal part of the shoe, the plough share (*phal*) is set in a groove.

The beam (*har*), inserted into the rectangular hole of the body, is a long, flat and slender piece of wood. It has two notches, on its inner side, where it is to be fastened with the yoke.

The yoke (*jual*) is curved out of a solid piece of wood. Near the two ends, there are two wooden pegs (*kul*) inserted through two holes. These pegs are meant to keep the bullocks in proper position.

The villagers themselves make their own plough excepting the blade which is made by the village blacksmith.

Basketry:

Baskets prepared by the Mawasi people are used in their day to day work which is netting of bamboo strips. *Tokni/baiya*, *dhutu* (for keeping fish), *tatta* (used for veranda cover), *goda* (use for hens) used for domestic work. *Baiya* is a big sized grain container and *tokni* is the smaller one. Sometimes netting of bamboo strips is a part of cart. It is used to sit upon and to carry many things by putting on it. There are two types of design of bamboo work found in the villages one is *check* and another is *twined*. *Check*, in which the warp and weft pass over and under each other singly, as in woven cloth. *Twined*, when two or more wefts, pass alternately in front and behind each of the warps, crossing them obliquely. *Kolari*, a long bamboo stick used in agricultural work. All these articles prepared from bamboo are useful in day to day work of the villagers.

Musical Instruments

Setum (Dafli, wooden frame covered with goat skin), *dholak* (made of *bija* wood) and *kartal* (made of Kansa) are traditional instruments of the Mawasi. They play these during marriage, birth and on other festive occasions.

Things related to animals

Dhuna, *tappar* bell are tied got the cattle's neck. *Tapper* is made of wood and its size is bigger. *Dhuna* a bell made of iron or bell-metal. Due to this cattle's can not stray away.

Change is inevitable. Material aspect has much susceptibility for change due to various innovative processes which go side by side with the technological advancement. These influence the mode of life of the people to a large extent who are on the way to urbanization. The earthen pots, brass and bell-metal utensils have found a replacement in light and handy aluminium and stainless steel utensils. Thus, change is easily perceptible on the above-mentioned aspects of material culture.

DOCUMENTATION

Musical Instruments No. 1

Local Name: Dholok

English equivalent: Double membrane drum

Material used: Bija wood, skin and cord

Description: This is a double membraned hollow resonating wooden body. The body of the drum is made by scooping out a tree trunk and has a maximum diameter near the middle from where the diameter decreases gradually towards the open ends. The two circular openings at two ends of the drum are covered with two circular pieces of animal skin which is strained by means of rings slipped round the edge of the openings. A string is attached longitudinally for hanging the drum to the neck.

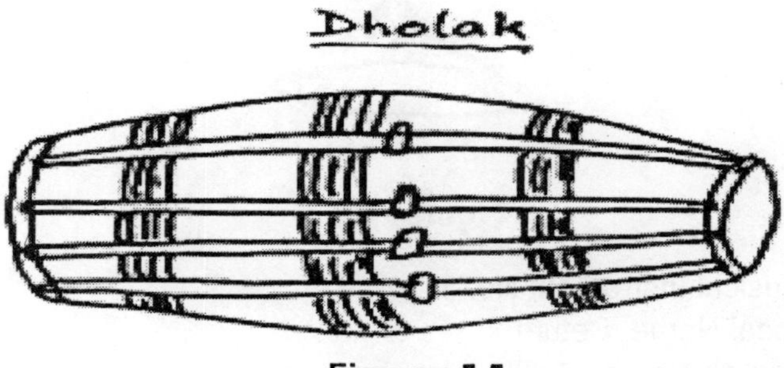

Figure : 5.5

Measurements: (a) Total length of the drum (A- Ai) – 20 inches (b) Diameter of the drum heads (Ai- Bi) & (A-B) -9 inches and 11 inches respectively.

Musical Instrument No.2 :

Local Name: Kartal

English equivalent: KHARTAL

Material used: Made of Kansa

Description: This is a plate like structure, in the middle of the plate is slight convex

Nature if use: Used during singing song.

Measurement :(a) Diameter of the Kartal (A- Ai) – 7 inches (b) (B-Bi) – 2 and half inches structure with a little hole where a cord can be put inside

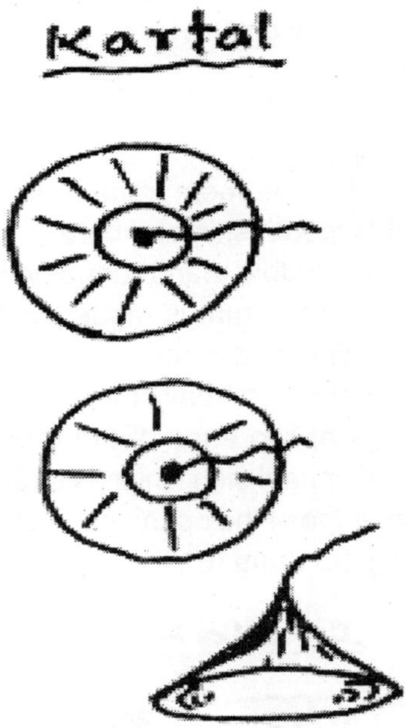

Figure : 5.6

Musical Instrument No. 3

Local Name: Setum

English Name: Dafli/ Drum single membraned

Material used: Wood, Goat skin and string
Description: This is a single membraned in shape. The body of the drum has been
from a block of the wood. A piece of skin is stretched over the one side opening of the opening body.
Measurement: (a) Diameter of the drum (A-Ai) - 6 inches
(b) Diameter of the skin (Ai-Bi) - 21unches
(c) Total length of the wooden drum -70 inches

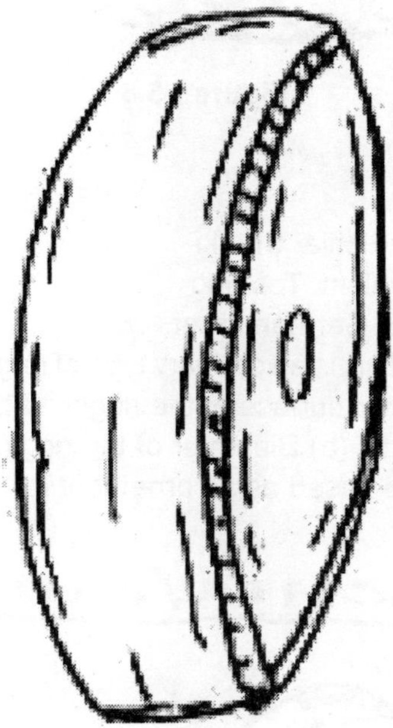

Figure : 5.7

Ornament No. 4
Local Name: Tora
English equivalent: Finger ring
Material used: German silver
Description: This is a small german silver ring with an inner flat surface and an outer rounded surface.
Measurements: Diameter of the ring (A-Ai) – 2.0 cms.
Nature of use: It is used as a finger ornament by females.

TORA

Figure : 5.8

Ornament No. 5
Local Name: Bichia/ Chutki
English equivalent: Toe ring
Material used: German silver
Description: This is an ordinary type of ring made of German silver. The inner surface of the ring is flat.
Measurements: (a) Diameter of the ring (A-Ai) – 2.2 cms.
Nature of use : used as an ornament for toe by females.

BICHIA/CHUTKI

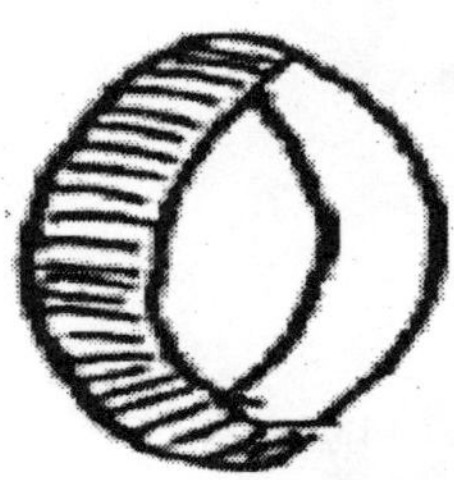

Figure : 5.9

Ornament No. 6
Local Name : Patti

English equivalent: Anklet
Material used: German silver
Description : This is a beat shaped solid and curved anklet made of german silver.
The inner surface is smooth and the outer surface has designs.
Measurements: (a) Length of anklet (A-Ai)-5 inch.
(b) Diameter of the body of the anklet (B-Bi) - 0.5 inch.
Nature of use: Used as an ornament for leg.
Method of use: It is slipped on the ankle.

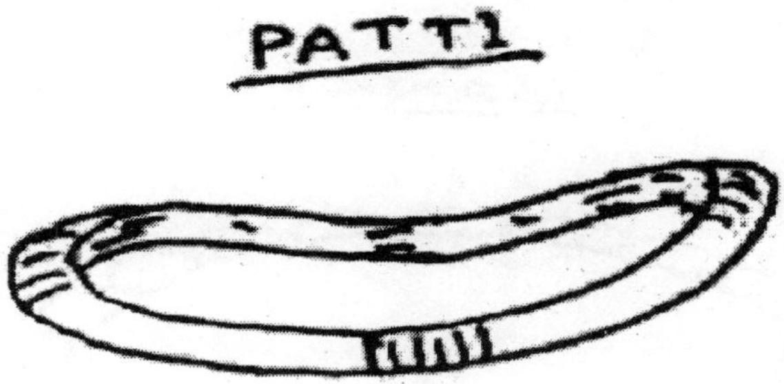

Figure : 5.10

Ornament No. 7
Local Name: Jhumka
English equivalent: Ear ring
Material used: Silver
Description: The ear ring has a 'U' shaped pendant and a U shaped hook arrangement. The surface of the pendant is wavy and covered with designs. The hook arrangement consists of an inverted U shaped structure having one curved hook at one end and the other end is turned into a ring which is fixed to one ring of the pendant.
Measurement: Distance from (A-Ai)- 1.2 inches.
Nature of use: Used as an ear ornament.

Method of use: The hook is passed through the hole on the ear lobe and fixed to the loop of the pendant.

Ornament No. 8
Local Name: Bangdi
English equivalent: Bangle
Material used: German silver

Description: This is a german silver bangle of circular shape. The bangle is of equal thickness throughout. The inner surface of the bangle is flat and outer surface is wavy and convex at regular intervals.

Measurements: Diameter of the bangle (A-Ai) – 3 inches.
Nature of use: Used as an ornament or wrist.

BANGDI

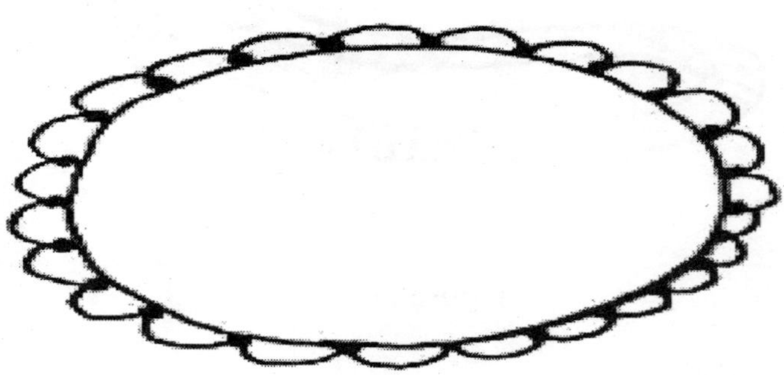

Figure : 5.11

Article No. 9
Local Name: Palai
English equivalent: Basket for keeping grains.
Material: Bamboo strips

Description: This is a globular basket made of thin and narrow bamboo strips in Check pattern. It has a flat circular base, rounded body and narrow mouth.

Measurements: (a) Total height of the basket (A-Ai) - 24 inches.

(b) Diameter of the mouth (A-B)-12 inches

(c) Circumstances of the basket (C-Ci) - 2"

Nature of use: For keeping grains in the house.

Material used: Manufactured by local villagers from bamboo strips collected from forest.

PALAI

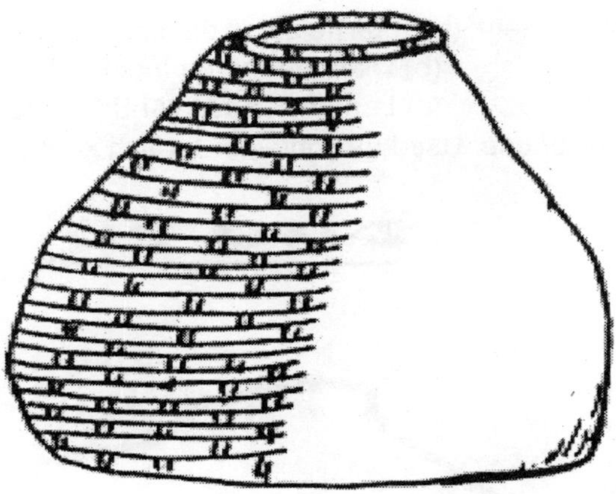

Figure : 5.12

Article No.10.

Local Name: Churi

English equivalent: Knife

Material used: Iron for blade and wood for grip

Maker place: Made by Lohar of local market.

Description: The knife consists of a blade and a grip. The blade is triangular in shape and has one sharp edge which is gently convex with a pointed tip. The other end is blunt, thick and concave. The blade has a shoulder at the posterior end followed by a broad and stout tamng. The tang is introduced into the handle. The wooden handle is a solid, short, piece of wood.

Measurements: (a) Length of the blade (A-Ai): 8 inches

(b) Maximum width of the blade: 1.5 inches

(c) Length of the handle (Ai-B) 5 inches

Nature of use: Used for cutting fine strips from bamboo.

Article No.11

Local Name: Fish Basket

Material: Bamboo strips

Description: This is a bamboo basket used for keeping fish. It has a small circular Mouth and a square base. The basket is woven in check pattern of Wicker work and has a bulging body. The basket is provided with a Handle made of cord tied like a loop over the mouth.

Measurement: (a) Total height of the basket (A-Ai): 8 inches
(b) Diameter of the mouth (A-B): 6 inches
(c) Length of base (Ai-Bi): 4 inches

Nature of use: Used as container for fish kept into it.

DHUTU

Figure : 5.13

Article No.12

Local Name: Supra

English equivalent: Winnowing Fan

Material use: Bamboo

Description: The specimen is 'U' shaped with posterior part slightly broader than the anterior part. It has a raised rounded posterior end and a straight anterior end. The body of the winnowing fan is made of bamboo strips in twin pattern. The margin is strengthened by tying bamboo splints all around.

Measurements: (a) Length of the straight (A-Ai): 1 ft 6 inches

(b) Breadth of the winnowing fan (A-B): 1 ft 3 inches

(c) Breadth of the posterior end (Ai-Bi): 1 ft.

Maker and Material use: Made by local basket makers by bamboo strips.

Nature of use: Used during separating dusts and chaffs from grains.

SUPRA

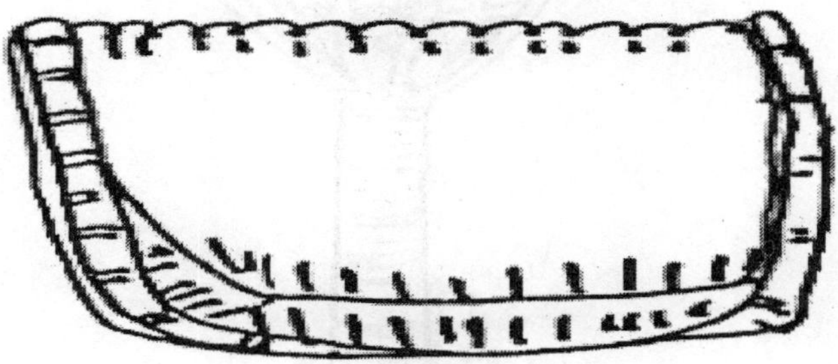

Figure : 5.14

Article No.13

Local Name: Goda

English Name: Hen Roost

Material use: Bamboo

Description: This is a basket with a stand made of same piece of bamboo. The anterior part of the bamboo is split into several rays and they act as warps. The warps are then woven with fresh wefts in check pattern. The posterior unsplit part serves as stand. A node in between the stand and the basket prevents further splitting. The basket is funnel shaped conical. The open mouth is wide and circular. The stand is cylindrical and straight.

Measurements: (a) Height of the stand (A-Ai) : 42 Inches (3 ft 6 in.)

(b) Diameter of the same: 5 inches

(c) Height of the basket: (A-B): 20 inches (1 ft 8 in.)

(d) Diameter at its mouth: (B-Bi): 36 inches (3 ft)

Nature of use: Used as hen roost and keep in the varanda (*osari*) of the house.

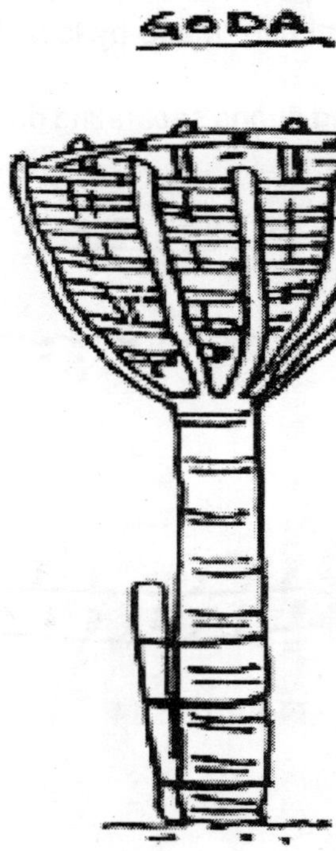

Figure : 5.15

Article No. 14

Local Name: Tappar

English equivalent: Wooden Bell

Material: Wood and fiber string

Description: The specimen consists of the following parts (i) the hollow wooden body (ii) one wooden clapper. The oval shaped body was fashioned from a solid block of wood with the help of a chisel and hammer. The clapper (wooden) is suspended into the hollow cavity by means of strings.

Measurements :(a) Maximum length of the wooden body(B-Bi) : 7 inches

(b) Maximum width of the same (A-Ai): 6 inches

(c) Length of the clapper (C-Ci): 3 inches

Nature of use; It is tied to the neck of the cattle.

Method of use: It is suspended to the neck of the cow or bullock. As the object moves, the clappers strike each other and produce a sound. To find out a cow into the jungle, this bell helps a lot as the sound gives an idea of the cattle's where about.

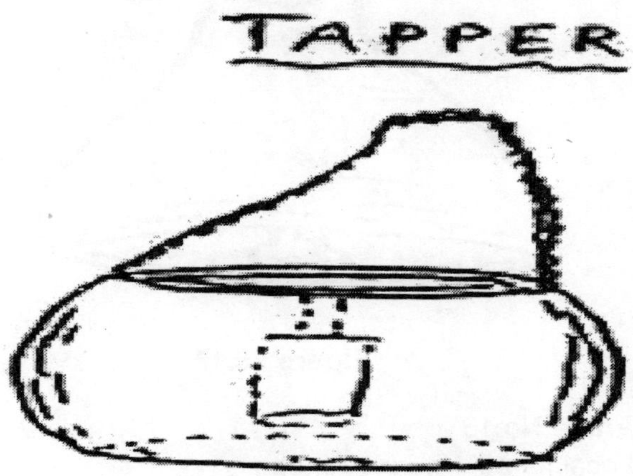

Figure : 5.16

Article No. 15

Local Name: Dhuna

English equivalent: Bell

Material: Iron

Description: The specimen consists of two identical bells suspended from a cord. A hollow funnel shaped gong which is widely opened below, and a solid club shaped striker which is suspended in the gong from a hook at the top. At the top of the bell, a circular aperture is found through which the cord carrying the bell is passed.

Measurement: (a) Height of the bell (A-Ai): 3 inch

(b) Diameter of the open end (Ai- B): 1 ft 5 in.

Nature of use: Used for hanging round the neck of goat or some other animals.

Maker: Purchased from a local market @Rs. 30/- to Rs.50/- according to size.

Figure : 5.17

Article No.16

Local name: Khurpi / Khurpa

English equivalent: Garden spade

Material: Iron for blade and wood for handle.

Description: The implement can be mainly divided into two parts (i) a blade and (ii) a handle. The handle is simple piece of wood, straight and uniform in diameter. One end of the handle is free and other end receives the tang of the blade. It is cylindrical throughout. The blade can be divided into following parts: tang, twisted neck, shoulder and body with working edge.The neck is twisted.

Measurements: Khurpi (a) (A-Ai) : 2.5 inches (b) (A-B): 3in. (c) (A-C): 2in. (d) (B-Bi):1.5in. (e) C-Ci): 1.5 in. (f) Length of the handle: (D-Di) -9.5 inch (g) Diameter of the handle- 1 inch Kurpa: (a) Length of the blade (A-B)-3.5 inch. (b) Breadth of the blade (A-Ai)- 2 in. (c) B-C: 1.2 in. (d) Length of the handle (D-Di):10 inch.

(e) Diameter of the handle: 3 inches.

Nature of use: Used for digging and weeding.

Material used: Local Lohar made the iron blade and handle made by the forest wood.

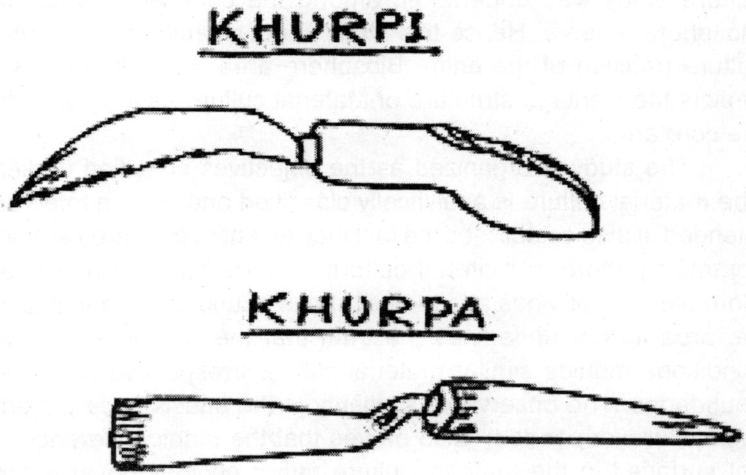

Figure : 5.18

Article No. 17

Local Name: Datre (Hasia)

English equivalent: Sickle

Material: Iron for blade and bamboo for handle (Muthia)

Description: It has two parts (i) a blade and (ii) a handle. The blade is crescent shaped with One concave toothed cutting edge; The opposite edge is thick, blunt and convex. The tip of the blade is pointed. Posteriorly the blade has a tang which is Introduced into a handle. The handle is uniform in girth and straight.

Measurements: (a) Length of the blade (A-Ai): 12 inch. (b) Maximum width of the blade (B-Bi): 1.5 inch, (c) Length of the handle.

Conclusion

The present endeavour aimed to document the material culture prevalent in the villages found in the core area of Pachmarhi

Biosphere Reserve. The study was conducted as part of the overall project entitled "Pachmarhi Biosphere reserve study". The material culture study was undertaken among the core villages of the Biosphere reserve. Hence this study is not claiming the Material culture tradition of the entire Biosphere area, while it obviously depicts the Heritage structure of Material culture as it belongs to the core area.

The study is organized as the objectives specified earlier. The material culture is analytically classified and documented as intended. It also establishes the fact that the entire core area shares common pattern in material culture in its production and use. Moreover, the obvious absence of ethnic bound craftsmanship in the area further underlines the truth that the similar ecological conditions moulds similar material culture irrespective of ethnic boundaries. The observable similarity in the subsistence pattern (agricultural depended) also proves that the ethnic difference is not surfaced in the material culture rather ecology governs the character of material culture. It is significant to note that gender plays a substantial role in material culture practice as it is observed that some of the items are marked by gender in use irrespective ethnic differences.

References

Census of India 1981 Catalogue of Tribal specimens, office of the Registrar General of India, Ministry of Human Affairs, NewDelhi.

Director, Anthropological 1961 "Peasant Life in India- A study Survey of India in Indian unity And Diversity", Anthropological Survey of India, Calcutta.

A Committee of the Royal 1967 Notes and Queries on Anthropological Institute Anthropology, Sixth edition, Routledge and Kegan Paul ltd, London

Hutton, J.H. 1944 Caste in India its Nature Function and its Origin, Oxford Univ. Press, Delhi.

Singh, K.S. 1994 The People of India, National Series, Vol. III, Anthropological Survey of India, Oxford Univ. Press, Delhi.

Tiwari, S.K. 1997 Wild Life Sanctuaries of Madhya Pradesh, APH Publishing Corporation, New Delhi.

Maity, Tamal 2001 "Ethnic Identity : A Case Study of Mawasi of Madhya Pradesh", published in Journal of Anthropological Survey of India, special issue "Ethnic situations in Central India", 2001 Vol. 50, No. 3, pp. 73-77.

6

Ethnic Composition of the Buffer Zone of Pachmarhi Biosphere Reserve : An Antropological Appraisal

Mukesh Sahay

Sheela Dasgupata

Introduction

Man has been living in the forest for ages to make a living from them. He was always in the forest with his bows and arrows. He used to take his traditional brew after worshipping his gods in the forest. Forest was the be all and end all of his life. They were also expert in protecting the forest with their ancient tradition and wisdom in order to keep natural ecological balance intact. The result was extensive tree cover and greenery even at the peak of the summer. This remarkable image of the forest was a common place over the past several decades in India.

But now the picture is totally changed in the Indian landscape over the past few decades. It is a landscape in which the natural world is continually being replaced by a world of artifacts: where trees, shrubs and grasses are giving way to plantations and crop fields, roads and buildings. A walking tour through the forest can help us understand the damage of natural ecological balance due to the utilization of natural resources. The pressure of population and growing demand for resources and poverty, which depend directly on natural resources for their survival, has also taken a heavy toll of the environment and also changed the social fabric.

To cope up this problem the anthropologist who are pioneer in studying man in biology and social structure added a new dimension to study natural eco systems in their discipline. Although, the trend over the last few decades shows that the name and concepts of ecology have become increasingly popular in Anthropology. A little consideration of the bio-environment as factor in culture form can be found in American Anthropology from 1920 to 1950. Several British Anthropologist of this period notably Childe (1936), Forde (1954) and Richards (1939) also got exposed to American influence and investigated the interaction of the environment with human culture.

So, there was a shift from the exclusive tribal studies to the bio-environment and Anthropologist embarked on an attempt to study biosphere reserves as it was centered around the ideas :

(i) Biodiversity must be saved within a matrix of human settlement – Thus, the man in its title.

(ii) Conservation should focus on entire ecosystem, not just single species.

And accordingly Pachmarhi was selected which was designated as biosphere reserve. Pachmarhi the vast, seemingly impenetrable forest of central India has long represented the epitome of virgin, chaotic nature and inhabited by indigenous people. In recent decades, however, the entire area has changed a lot and has presented a completely different picture. Since long the nearby area of Pachmarhi forest have been populated by many different cultures and societies that have left their marks on the landscape.

The present paper on Pachmarhi biosphere reserve is focused primarily on the population structure and ethnic pattern of that area.

Pachmarhi was designated as Biosphere reserve by the government of India on 3rd March, 1999. it lies between latitude 20° 10′, and 22° 50′ N 77° 45′, and 78° 56′ E longitude. It is situated in central Narmada Valley and extends over the Satpura Hill ranges.

The name Pachmarhi is derived from "Panch – Mattu" or five huts from ancient caves which have been carved out in a small hill rising abruptly from an open part of the country. Pachmarhi envelopes three wild life conservation units namely:

(i) Bori Sanctuary
(ii) Pachmarhi Sanctuary
(iii) Satpura National Park

Out of these three, the Satpura National Park comprises the Core Zone. The remaining two i.e. Bori and Pachmarhi sanctuary surrounding the core zone have been designated as buffer zone.

Pachmarhi sanctuary derives its name from Pachmarhi town, the only hill station of central India. It was bifurcated from Bori sanctuary in the year 1977 but again in 1981 certain area of this sanctuary was merged with Satpura National Park. So, now Pachmarhi sanctuary has an area of 417 sq. km. This sanctuary falls under the jurisdiction of Pipariya and Sohagpur tehsil of Hoshangabad district.

Bori sanctuary is the oldest reserve forest in India. The sanctuary extends over an area of 485.72 Sq. Km., slightly bigger than Pachmarhi sanctuary. It comprises of practically whole of Bori forest range and part of Sukhtawa forest range. It has been declared as a wild life sanctuary under the Wild Life Protection Act 1972 for the first time in 1975 and subsequent in 1977. Bori sanctuary falls under the jurisdiction of Sohagpur Tehsil of Hoshangabad district.

The basic information has been collected from EPCO (Environmental Planning and Co-ordination Organization), which is the nodal agency on behalf of Government of Madhya Pradesh for Pachmarhi biosphere reserve. In addition to EPCO, district officials of all the three districts, village Patwaris, local people, forest officials were also contacted for collecting data. All these informations have been further supplemented by census publications particularly district census hand books of the year 1961, 1971, census reports of 1991 and 2001.

The Buffer Zone

The buffer zone which surrounds the core zone has an area of 4927.28 Sq. Kms. having 528 villages distributed in to entire stretch of the buffer zone. It is spread over three civil districts i.e. Betul, Chindwara and Hoshangabad. The reason for inclusion of certain areas of the above mentioned districts may be the following.

(a) Culturally, socially and ecologically these areas have a close resemblance.

(b) Their traditional natural resources are also more or less one and the same and equally distributed in these areas.

(c) The strategies of the consumption and management of natural resources of the local people are also similar in many ways.

(d) To restrict the activities of the local people of the area as well as the outsiders within this area in order to deduce its effects on core zone.

The district-wise break up of the area of buffer zone is as follows:

Betul	561.13 sq. km.
Chindwara	1386.56 sq. km.
Hoshangabad	2971.59 sq. km.

Hoshangabad comprises a major part of this area owing to wide extension in all most all the tehsils of the district but Betul and Chindwara districts have less contribution due to smaller extent as only two tehsils of each districts have been included.

Name and Nos. of Tehsils of Districts in Buffer Zone

Sl. No.	*District*	*Tehsil & Block*
01.	Betul	(a) Ghoradongri (b) Shahpur
02.	Chindwara	(a) Jamai or Junnardev (b) Tamia
03.	Hoshangabad	(a) Babai (b) Bankhedi (c) Hoshangabad (d) Pipariya (e) Itarsi – Kesla (f) Sohagpur

All the villages included in the above mentioned tehsils are part of buffer zone.

Boundary of Biosphere Reserve

Buffer zone of Pachmarhi biosphere reserve has a boundary extended in all the three districts and starting from the north-west corner near Pathrota on Itarsi – Nagpur state highway No.22 and moving clockwise.

North – The road along north side of the Tawa left bank canal north of Pathrota at the point where it meets the state highway eastward to Kandaikalan, then northward along pucca road to Gurra railway station, again northwards along pucca road to Rampur, to Tama river and across it, and along pucca road to Babai, then all along state highway eastwards via Sohagpur, Pipariya and Bankhedi up to Dudhi river.

East – Along the east bank (right bank) of Dudhi river, southwards (up stream) up to Mukunda village up to which the river is the inter-district boundary between Hoshangabad and Narsinghpur and Chindwara and Narshingpur districts, then southwards (up stream) in to Chindwara district via Pindrai, Kosmi to Nagdaun along the same river, then along the western boundary of Haildubbha reserve forest of Chindwara district up to Sidhauli on Tamia plateau and from Sidhauli along pucca road up to Bijora, then to Tamia, from Tamia along kachha road via Jhil Pipariya to Tamai.

South – From Tamai in Chindwara district westwards to Dongariya, Damua then northwards along Kanhan river and fair weather road to Bharuakhapa, tarai then westwards along foot path to Matiya, on the boundary between Chindwara, Betul, then along foot path to Imlikheda in Betul, then along cart track to Nishchitpur, to Chopra and along pucca road westwards up to state highway No. 23 at Shahpur.

West – From Shahpur in Betul district along state highway No. 23 northwards via Bhaunra, Sukhtawa, Kesla up to Pathrota and crossing of Tawa left bank canal.

Distribution of Village in Buffer Zone

The buffer zone has so many villages. Agriculture is the main stay of the population of these villages as well as the main source of income. Working in the forest area is the supplementary source of income for the villagers of buffer zone as forest land is as much as 62.47% in this area, while agriculture land is 29.38%. The EPCO, which is the nodal agency for Pachmarhi biosphere reserve had designated 583 villages in buffer zone area, total 508 villages belong to revenue villages rest 67 villages are forest villages. These villages have been distributed in all the three districts earmarked for buffer zone. But our study reveals that 528 villages have been distributed in buffer zone of which 465 belong to revenue villages and district.

Hoshangabad shares 53.47%, Chindwara 36.77% and Betul is the lowest having 10.75% villages. There are also 54 forest villages in buffer zone which is shared by Hoshangabad 77.77% and Betul 22.42%. This shows a shortfall of 47 villages in this area. The difference in this number is caused on account of many reasons of which the most important are:

(1) Counting a hamlet of the village as a separate village. Most of the villages of buffer zone are constituted with one or two hamlets which exists apart with a name from the main village and known as *dhana* or *tola* of that village. This led them to believe existence of another village and listed them as separate village.

(2) In some cases repetition of the name of the villages in the list of EPCO have been observed by the authors. This also may be one of the reasons for discrepancy in their list.

(3) Authors have also observed that few villages which are existing outside the buffer zone have been included in the list of EPCO. Due to this reason the number of villages in their list has gone up.

We have observed this shortfall of the villages in all the three districts. Only in case of Chindwara district regarding forest villages we are at par with EPCO. As far as the urban agglomeration is concerned the difference is not much, rather in Hoshangabad and Chindwara our study is evenly balanced with EPCO. Only in Betul there is one addition of urban agglomeration. Another most remarkable aspect is inclusion of abandoned villages in buffer zone. The authors have identified altogether 32 villages in this area which are uninhabited. There is a substantive number of abandoned villages in all the three districts.

Table 6.1
Distribution of Villages in the Buffer Zone

	Districts			
	Hoshangabad	*Betul*	*Chindwara*	*Total*
Total Nos. of revenue villages	269	51	188	**508**
	244	50	171	**465**
Total Nos. of forest villages	49	18	00	**67**
	42	12	00	**54**
Total Nos. of urban Agglomeration	06	00	02	**08**
	06	01	02	**09**
Total settlement	325	69	190	**583**
	292	63	173	**528**
Abandoned villages	17	07	08	**32**

As per EPCO 08 villages of core area are also included.

All the 528 villages are located in ten tehsils/blocks of three districts of buffer zone. The tehsil/block-wise list of villages falling in buffer zone as per EPCO and our study is given below. The Table 6.2 shows the existence of settlements in all the ten tehsils/ block of all the districts of buffer zone. According to this table Tamia block of Chindwara district has got highest number of revenue villages i.e. 125, whereas Pipariya tehsil of Hoshangabad is second highest comprising 81 villages but Shahpur in Betul district has only 23 villages.

Table 6.2
Block wise Revenue, Urban and Forest Settlements in Pachmarhi Biosphere (As Per EPCO)

Districts	*Block/Tehsil*	*Number of Settlements*			
		Revenue	*Urban*	*Forest Villages*	*Total*
Hoshangabad	Itarsi	42	02	15	59
	Hoshangabad	05	00	00	05
	Babai	24	01	00	25
	Sohagpur	66	01	24	91
	Pipariya	81	02	10	93
	Bankhedi	51	00	00	51
	TOTAL	**269**	**06**	**49**	**324**
Betul	Ghoradongri	28	00	17	45
	Shahpur	23	00	01	24
	TOTAL	**51**	**00**	**18**	**69**
Chindwara	Tamia	125	00	00	125
	Jamai	63	02	00	65
	TOTAL	**188**	**02**	**00**	**190**
	GRAND TOTAL	**508**	**08**	**67**	**583**

(Source: Census of India, 2001)

As stated 528 villages have been distributed in an unequal number in three districts of buffer zone. The number of villages vary due to the size of the area included in the buffer zone. Since a large area of Hoshangabad district is scattered in six tehsils/ block have been taken as part of buffer zone. So, the maximum number of villages have been included in this district whose number is 292 including 06 urban settlements. Out of this 292 villages, 244 villages are part of revenue villages and 42 are forest villages.

Only 06 are urban settlements in Hoshangabad. The buffer zone also comprises two sanctuaries i.e. Pachmarhi and Bori. In these two sanctuaries only Pachmarhi has both revenue as well as forest villages whereas, Bori sanctuary has only forest villages. All the villages (both revenue and forest are part and parcel of buffer zone).

Table 6.3
Block-wise Revenue, Urban and Forest settlements in Pachmarhi Biosphere
As per present study 2005

Districts	*Block/Tehsil*	*Number of Settlements*				*Abandoned village*
		Revenue	*Urban*	*Forest Villages*	*Total*	
Hoshangabad	Kesla (Itarsi)	39	02	11	52	07
	Hoshangabad	05	00	00	05	00
	Babai	23	01	00	24	01
	Sohagpur	57	01	22	80	03
	Pipariya	71	02	09	82	03
	Bankhedi	49	00	00	49	03
	TOTAL	**244**	**06**	**42**	**292**	**17**
Betul	Ghoradongri	28	00	11	39	06
	Shahpur	22	01	01	24	01
	TOTAL	**50**	**01**	**12**	**63**	**07**
Chindwara	Tamia	112	00	00	112	07
	Jamai	59	02	00	61	01
	TOTAL	**171**	**02**	**00**	**173**	**08**
	GRAND TOTAL	**465**	**09**	**54**	**528**	**32**

Betul district is contributing less area in the buffer zone. The number of villages in this part of buffer zone has gone down substantially as only two tehsils of the district i.e. Ghoradongri and Shahpur are within the buffer zone and together contributing only 63 villages in which 12 are forest villages of Ghoradongri tehsil and only one urban agglomeration as Shahpur. On the contrary, Chindwara district too has only two tehsils in buffer zone but the area of thse two tehsils are much bigger than Betul. So, this part of buffer zone has got more number of villages. Like villages, the buffer zone has many townships and urban agglomerations.

Hoshangabad district covers the largest area and within this area a number of townships like Pipariya, Pachmarhi, Pathrota, Sohagpur, Babai and Ranipura (Tawa Nagar) are situated. These townships are now part of buffer zone. Similarly Betul and

Chindwara too have a number of townships and human settlements but among those only Shahpur in Betul, Jamai or Junnardeo and Damua in Chindwara falls under buffer zone. So, altogether nine urban settlements or urban agglomerations are located in buffer zone. Here also the authors have added one more urban centre in the studied list which is in opposition of EPCO because they have shown only eight numbers in their list. All these townships of buffer zone are inhabited by a number of communities including those who are categorized as SC and ST.

Demographic and Ethnic Pattern of Buffer Zone

Buffer zone consists of cluster of villages as well as several townships. It falls within the jurisdiction of three districts of Madhya Pradesh. As a result the buffer zone provides a picture of much greater heterogeneity in its population structure, but enjoying both social and cultural continuity and united by a common dialect known as Bundelkhandi.

As per 1991 census, the total population of buffer zone is about 4,07,504 of which 67.67% of the population falls in revenue villages, 29.35% in urban areas and rest 1.98% in the forest villages. The highest population contribution is of Pipariya, Jamai and Sohagpur tehsils having percentage of 20.39%, 17.65% and 15.13% respectively of the total buffer zone population. Forest village population is maximum in Ghoradongri block of Betul district. Chindwara and Betul districts are the most prominent trial belt as scheduled tribe population constitute 48.95% and 40.04% respectively of the total population. Tamia block of Chindwara district has the largest scheduled tribe population, as about 2/3rd of this population is tribal.

Contrary to the 1991 census, 2001 census shows the population of buffer zone as 4,99,135, which reveals an increase in the population by 91,631 persons. So, 22.5% increase in the total population of the entire buffer zone.

Obviously, a large chunk of population, 64.43% is coming from Hoshangabad district owing to largest number of villages but Betul has less contribution of 9.6% as this district is adding only 50 villages in the buffer zone. Chindwara is adding 26.4% due to the inclusion of 173 villages.

Table 6.4
Total Number of Population in Buffer Zone

Tehsils	*Districts*			*Grand Total*
	Betul	*Chindwara*	*Hoshangabad*	
Babai	00	00	40,799	40,799
Bankhedi	00	00	49,177	49,177
Block – HB	00	00	4,273	4,273
Sohagpur	00	00	73,803	73,803
Itarsi – KE	00	00	49,228	49,228
Pipariya	00	00	1,04,332	1,04,332
Ghoradongri	21,926	00	00	21,926
Shahpur	23,809	00	00	23,809
Jamai	00	75,060	00	75,060
Tamia	00	56,728	00	56,728
Grand Total	**45,735**	**1,31,788**	**3,21,612**	**4,99,135**

(Source: Census of India, 2001)

Table 6.5
Total Number of Male Population in Buffer Zone

Tehsils	*Districts*			*Grand Total*
	Betul	*Chindwara*	*Hoshangabad*	
Babai	00	00	21,373	21,373
Bankhedi	00	00	25,690	25,690
Block – HB	00	00	2,323	2,323
Sohagpur	00	00	38,781	38,781
Itarsi – KE	00	00	25,635	25,635
Pipariya	00	00	55,494	55,494
Ghoradongri	11,201	00	00	11,201
Shahpur	12,189	00	00	12,189
Jamai	00	38,490	00	38,490
Tamia	00	29,198	00	29,198
Grand Total	**23,390**	**67,688**	**1,69,296**	**2,60,374**

(Source: Census of India, 2001)

The male population of entire buffer zone is 260374. Hoshangabad is contributing the bulk of male population having 65.2%, Chindwara is 26% and Betul only 11.69%. The reason for this maximum and minimum of the male population within all the three districts of buffer zone has already been stated above.

The female population of entire buffer zone is 238762. Singnificantly the female population is less than their counterpart and the shortfall is 21612. Here also Hoshangabad has outnumbered as it includes 152327, female having 45.34%, followed by Chindwara which has 64100 females with 42.98% and lastly Betul with only 22335 females and the percentage is only 11.69%.

Table 6.6
Total Number of Female Population in Buffer Zone

Tehsil	*District*			*Grand Total*	*Sex Ratio*
	Betul	*Chindwara*	*Hoshangabad*		
Babai	00	00	19426	19426	908
Bankhedi	00	00	23484	23484	914
Block – HB	00	00	1950	1950	839
Sohagpur	00	00	35022	35022	903
Itarsi – KE	00	00	23615	23615	921
Pipariya	00	00	48830	48830	879
Ghoradongri	10715	00	00	10715	956
Shahpur	11620	00	00	11620	863
Jamai	00	36570	00	36570	950
Tamia	00	27530	00	27530	942
Grand Total	**22335**	**64100**	**152327**	**238762**	**916**

(Source: Census of India, 2001)

From the Tables 6.5 - 6.6 of male and female population it is clear that females are less in number than males in every part of buffer zone from Chindwara to Hoshangabad but in Betul district the difference of male female population is not as much as figured in other two districts of buffer zone. In other words it can be said that the quantity of decline in the male female population is not uniform throughout the buffer zone and varied considerably from one district to the other. Hence, Betul shows the highest sex ratio which is 954, then comes Chindwara where the sex ratio is 946 and the lowest 899 is in Hoshangabad. The over all sex ratio of entire buffer zone is 916. Malnutrition of the female child is perhaps the main reason of gradual decline in the sex ratio of buffer zone.

The number of household in three districts of buffer zone viz. Hoshangabad, Chindwara and Betul is 60852, 25283 and 8286 respectively.

Table 6.7
Total Nos. of Households in Buffer Zone Village

Tehsils	*Districts*			*Grand Total*
	Betul	*Chindwara*	*Hoshangabad*	
Babai	00	00	7,883	7,883
Bankhedi	00	00	9,212	9,212
Block – HB	00	00	772	772
Sohagpur	00	00	13,972	13,972
Itarsi – KE	00	00	9,206	9,206
Pipariya	00	00	19,807	19,807
Ghoradongri	3,864	00	00	3,864
Shahpur	4,422	00	00	4,422
Jamai	00	14,207	00	14,207
Tamia	00	11,076	00	11,076
Grand Total	**8,286**	**25,283**	**60,852**	**94,296**

(Source: Census of India, 2001)

When the numbers are combined together then the figure is 94296 forming the total number of households is in the entire buffer zone. The authors could not ascertain the actual number of

Table 6.8
Average Household size (Persons per household) of Buffer Zone

Tehsils	*Districts*			*Grand Total*
	Betul	*Chindwara*	*Hoshangabad*	
Babai	00	00	5.64	5.64
Bankhedi	00	00	5.38	5.38
Block – HB	00	00	5.58	5.58
Sohagpur	00	00	6.16	6.16
Itarsi – KE	00	00	5.54	5.54
Pipariya	00	00	5.31	5.31
Ghoradongri	5.72	00	00	5.72
Shahpur	5.53	00	00	5.53
Jamai	00	5.47	00	5.47
Tamia	00	5.20	00	5.20
Grand Total	**5.65**	**5.29**	**5.63**	**5.52**

(Source: Census of India, 2001)

household possessed by the individual ethnic groups of the buffer zone. So, it is presumed that all the ethnic groups of the buffer zone are distributed within this household. The average household size among the population of buffer zone is 5.65% in Hoshangabad, 5.5% in Betul and 5.2% in Chindwara district respectively.

Ethnic Groups of Buffer Zone

Buffer zone is a vast area which is inhabited by a number of ethnic groups. Some of them are oldest inhabitants of the area but few are here in the surroundings since generations. They are also now considering themselves as the oldest inhabitants. Among them mostly religious groups such as Christians, Jains, Sikhs and Sindhis are included. The ethnic groups of buffer zone is divided into five categories as General Category, Religious Groups, Scheduled Castes, Scheduled Tribes and Other Backward Classes (OBC).

Table 6.9
Table Showing Ethnic Groups in the Buffer

General Category	*OBC*	*ST*	*SC*	*Religious Groups*
Brahman	Ahir	--	--	--
Rajput	Agarwal	--	--	--
Raghuvanshi	Bania	--	--	--
Bishnoi	Dhobi	--	Basor	Christian
Goswami	Kalar	Bharia	Chamar	Jain
Kayastha	Kerar	Gond	Katiya	Muslim including Bohra
Purbia	Kurmi	Korku	Kotwar	Sikh
--	Kachhi	Mawasi	Mehra	Sindhi
--	Kumhar	--	Bengali	--
--	Gujjar	--	--	--
--	Gadaria	--	--	--
--	Mali	--	--	--
--	Nai	--	--	--
--	Teli	--	--	--
07	**14**	**04**	**06**	**05**

So, there are about 31 ethnic groups and five religious groups are present in the buffer zone. After seeing the Table 6.9 on ethnic groups one is indeed overwhelmed by the ethnic heterogeneity of the area. But it soon becomes apparent that the bulk of the population is formed by a limited number of ethnic groups. Division wise upper castes group constitutes 22.5%, OBC - 45.2%, ST - 12.9% and SC – 19.3% of the total ethnic groups. Among ethnic groups OBC outnumbered all the categories of the groups and registered their presence in all the three districts of buffer zone. It is also evident that the Gond, Korku, Mawasi and Bharia constitute the major share of the total population.

Majority of the religious groups are scattered in different townships and urban agglomeration of the buffer zone. The other ethnic groups belonging to the Hindu and tribal are spread every where in the buffer zone.

Occupation of the Ethnic Groups

Agriculture is the traditional occupation for most of the ethnic groups such as Ahir, Kurmi, Kacchi, Kirar, Rajpur, Goswami, Raghuvanshi and Brahman. Kayasthas are exclusively a working class and mostly engaged in government, semi-government offices and other establishments. Muslims are also working in government and semi-government offices, but majority of Bohra Muslims are engaged in trade and business. Sikhs and Sindhis are mostly engaged in transport business and contractorship, whereas Aggrawal, Jain and Bania are petty shopkeepers and grocers.

Chamar, Kumhar, Dhobi, Nai, Lohar and Kotwar are service castes and still doing their traditional occupations wherever they are living in the buffer zone. In addition to their traditional occupation, a few of them have agricultural land, so they are also doing agriculture as a secondary occupation. Gond, Bharia, Korku and Mawasi are living in the villages and agriculture is their occupation as well as main source of income. Working in the forest area is the supplementary source of income of these people. Collection of fire wood from the forest and selling it in the nearest market area may also be taken as secondary occupation for them.

Distribution of Population in Buffer Zone

Both Pachmarhi and Bori sanctuaries form together a cluster of 52 villages. There are eight forest villages out of 35 villages in Pachmarhi sanctuary and unexpectedly all the 17 villages of Bori sanctuary are forest villages. So, combining together all the forest villages of both the sanctuaries, it reaches half way of the total number of villages. In other words it can be said that both the sanctuaries have almost 50% forest villages. In these villages only

Gond, Mawasi, Korku are living excepting three villages of Bori sanctuary where Gawali (OBC) are also residing with them. But surprisingly village Kanjighat in revenue circle of Pachmarhi sanctuary is a uniethnic village where only Bharia tribe is residing. Here the case of Bharia is unique as their population does not share a single habitation in any of the forest villages of the sanctuaries. However, they are widely distributed in other villages of Pachmarhi sanctuary.

The other 27 villages in revenue circle of Pachmarhi sanctuary vary widely in the number of ethnic groups. Out of 27 villages, 17 villages don't have any population of scheduled castes whereas 10 villages have only scheduled tribe population. Other ethnic groups of the 17 villages are Brahman, Gadaria, Kurmi, Mawasi and Bharia. There are 44 revenue villages and only one forest village in Pipariya block. In all these villages the population is a mixture of ethnic groups mentioned in the list excepting Gujjar and Gawali but the three dominant tribal groups namely Gond, Mawasi and Korku live in each and every village.

The Sohagpur segment is contributing to 57 villages and five forest villages in the buffer zone. This five forest villages are out side Bori sanctuary and inhabited only by Gonds. Among 57 villages of revenue circle, 11 villages are without scheduled castes population and five villages are without scheduled tribe population. Surprisingly the eleven villages where scheduled castes are absent, the population of scheduled tribe is in considerable number in addition to Brahman, Dhanuk, Gujjar, Muslim, Kachhi, Nai, Raghuvanshi, Purbia and Teli. The five villages without Scheduled Tribe population are having Brahmin, Kachhi, Goswami, Raghuvanshi in addition to Katia, Mehra, Kotwar who are categorized as scheduled castes.

Kesla – Itarsi area of buffer zone has altogether 50 villages, in which 11 are forest villages. Out of this 50 villages, nine villages (three from revenue circle and six from forest village) don't have scheduled castes population. However, other castes group such as Brahman, Kurmi, Rajput, Kachhi, Dhanuk, Gond, Korku are inhabiting in the revenue circle villages either in a sizable number or in a less number but in all the six forest villages only Gond and Korku are inhabiting. Among all the ethnic groups in this part of buffer zone, presence of Gond in each and every villages strengthens the widespread local tradition that Gonds used to be numerically dominant group in the area in th past. The rest 41 villages of the area have mixed population which may be the combination of three or four ethnic groups or more than that.

Babai, the smallest segment of the buffer zone has only 23 revenue villages in which three villages are without Scheduled Castes population and one village is without Scheduled Tribe population. There is only one village in this zone which doesnt have any population of ethnic groups categorized as Scheduled Castes, tribes and OBC. However, the general ethnic groups such as Brahman and Rajput are residing in this village. Ethnic groups like Gond, Kalar, Kachhi, Dhanuk, Gujjar are the inhabitants of those villages where Scheduled Castes population is not in existence and Chamar, Kalar, Mehra, Dhobi are the inhabitants of only the non-tribal villages of this zone. The rest 18 villages have mixed population but the Gonds are found in every village except few. The most striking feature of this zone is non-existence of Korku tribe in any of the villages.

Bankhedi is the third largest segment of buffer zone which shares 49 villages of revenue circle only. These villages also resemble the villages of other segments as regards the population structure is concerned like other segments of the buffer zone, this zone too has eight villages without the population of Scheduled Castes. However, population of other ethnic groups such as Ahir, Bharia, Gond, Gujjar, Kalar, Teli, Raghuvanshi are inhabiting in these villages. Similarly, three villages of this zone don't have any Scheduled Tribe population but Basor, Chamar, Katia, Kalar, Raghuvanshi, Teli are living here. Six villages of this zone have both Bharia and Gond population in addition to the other ethnic groups. This Gond – Bharia combination resembles the villages of Pachmarhi sanctuary where such type of combination is also existing in more than 20 villages and strengthens the view that like Gond, Bharia are also the oldest settlers of the area. The population of Ahir is sufficient in this part of the buffer zone as they are found in 15 villages. Probably they are more in numbers in this segment as compared to the other segments like Sohagpur and Itarsi–Kesla where they are distributed only in few villages. Rest

38 villages don't have any specific constitutional ethnic groups, rather inhabited by three or four categories of the ethnic groups.

The Shahpur segment of Betul district which is the smallest area of buffer zone has 21 revenue villages and one forest village. The seven villages of this zone don't have any scheduled castes population but these villaghes have considerable number of Gond and Korku population with an exception of one village where Rajput is living with Gond and another, where Gayaki (a sub-group of Gond) is living with Korku. The only forest village of this zone is inhabited by Gond, Korku and Chamar. The rest 14 villages have mixed population of Brahman, Rajput, Katia, Mehra, Korku, Ahir, Chamar etc, but the most remarkable aspect is appearance of Gond in all these villages either as a major group or a smaller group.

There are 28 villages and 11 forest villages in Ghoradongri. All the forest villages are inhabited by Gomd only with an exception of one village where Mehra is also living with Gond. Nine villages of this zone are without Scheduled Castes population where a mixture of other ethnic groups such as Gond, Korku, Rajput, Teli, Gawali, Brahman are living. Eight villages are exclusively inhabited by Bengali refugees who came to these villages during 1964 to 1981 in batches from different refugee camps of Chhalisgarh and Gondia in Maharashtra. The rest 11 villages of this zone have mixed population.

Jamai also known as Junnardeo is in Chindwara district and contributing to 59 villages of revenue circle only. The ethnic groups of these villages are not different from other villages of the buffer zone, which is under Hoshangabad district. The distinguishing feature is the non-existence of Scheduled Castes population in 27 villages, which constitutes nearly half of the total villages of this zone. Out of 27 villages without Scheduled Castes population, seven villages are inhabited by Gond and Mawasi, five villages are inhabited only by Mawasi and in the rest 15 villages Ahir, Kalar, Gond, Mawasi and Teli are inhabiting. There is only one village which is without Scheduled Tribe population and in this village Chamar is living with Ahir and Teli. Bharia tribe also made their appearance in two villages of this zone along with Gond and Kalar. Both Scheduled Castes and Scheduled Tribes population are found only in four villages and they are Chamar, Katia, Gond and Mawasi, simultaneously five villages are also having the same ethnic groups but Ahirs too are sporadically distributed in these villages. They are also inhabiting along with Mawasi in five villages. The rest 15 villages are inhabited by mixed ethnic groups in a combination of Mawasi, Katia, Chamar or Mawasi, Kalar, Gond, Katia or Mehra, Chamar, Ahir, Teli.

Tamia is also in Chindwara district and largest segment in respect of the villages as it includes 112 villages of revenue circle. No other segment has occupied this much of villages in buffer zone. Absence of Mawasi tribe is the remarkable feature of this segment and presence of Gond in 93 villages (83%) is the most striking feature of this segement. Out of this 93 villages, 21 villages are uniethnic where no other than the Gond is living. 06 villages are also exclusively inhabited by Bharia tribe. 14 villages include Brahman in their population in addition to the other ethnic groups such as Ahir, Sutar (Badhai), Bania, Chamar, Lohar, Mehra and Rajput. Similarly Rajputs are also found in 13 villages along with either the above mentioned ethnic groups or few more than this. The Brahman and Rajput in proportion to the other ethnic groups of this segment show a distinct tendency and prefer to live in a larger heteroethnic villages. There are 21 biethnic villages where combination of Gond-Rajput, Brahman-Rajput, Gond-Teli, Gond-Mehra, Gond-Chamar, Gond-Ahir and Gond-Lohar are living. Chamar, Katia and Mehra are the three Scheduled Castes communities living in this segment of buffer zone. Among them Mehra is living along with other communities in 18 villages, Chamar and Katia are living alsong with other communities in 16 and 10 villages respectively, Ahir is also living in 19 villages along with communities such as Brahman, Rajput, Lohar, Chamar, Katia, Gond, Sutar, etc.

Urban Area of Buffer zone

There are 09 urban areas in buffer zone. The maximum number is in Hoshangabad i.. 06(66.7%), Chindwara 02 (22.2%) and only 01 (11.1%) in Betul district respectively. Among these urban areas Sohagpur is the oldest one as it was recognized as a

Table 6.10

Distribution of Population in Urban Areas of Buffer Zone

Districts of Buffer Zone	*Urban Areas in Buffer Zone*	*Total Population*					*Population of SC & ST*			
		Total	*Male*	*%*	*Female*	*%*	*SC*	*%*	*ST*	*%*
Hoshangabad	Pipariya	41317	21753	52.6	19564	47.3	5309	12.8	1571	3.8
	Pachmarhi	11370	6416	56.4	4954	43.5	2139	18.8	1261	11.1
	Sohagpur	22339	11677	52.2	10662	47.7	3381	15.13	668	3.1
	Babai	14591	7660	52.9	6931	47.5	1700	11.65	558	3.8
	Pathrota	6136	3211	52.3	2925	47.6	1221	19.9	725	11.8
	Ranipura Tawa Nagar	5041	2676	53.1	2365	46.9	1648	32.7	572	11.3
Chindwara	Jamai	22440	11605	51.7	10835	48.3	3711	16.6	352	1.6
	Damua	15857	8185	51.6	7672	48.3	4487	28.29	1800	11.3
Betul	Shahpur	3997	2112	52.9	1885	47.1	403	10.1	250	6.25

(Source: Census of India, 2001)

town since 1901, whereas Pachmarhi and Pipariya in 1931 and Babai in 1971. Pathrota, a village first time appeared as an urban outgrowth (OG) in 1971 census. Ranipura was a village but it was developed as a township for irrigation project on Tawa river during second five year plan and since then it is known as Tawa Nagar also. Now both Pathrota and Ranipura (Tawa Nagar) are recognized colony under Itrasi.

Shahpur, the only urban area of buffer zone in Betul district is categorized as census town but not much developed and looks like an urbanized village. Jamai or Junnardeo in Chindwara district is a pilgrimage centre for local as well as people of neighbouring areas. This area is fully urbanized with all civic amenities and categorized as Nagar Parishad (NP) whereas Damua is also under Chindwara district and urban area of buffer zone is categorized as census town without much civic amenities.

The urban areas of buffer zone is characterized by the less representation of the tribal population. Only in Pachmarhi, Pathrota, Ranipura and Damua their population is above 10%. Population wise Pipariya ranks first and Shahpur is at the lowest. The percentage of male population is highest (56.42%) in Pachmarhi and lowest in Damua (51.6%). Surprisingly the percentage of female population in all the three areas of buffer zone is below 50% mark. However, Damua and Jamai have slightly higher percentage (48.38% and 48.3%) of female population in comparison to the other 07 areas of the buffer zone. The population of Scheduled Castes is adequately placed in urban areas and Ranipura (Tawa Nagar) shows the highest percentage (32.7%) of Scheduled Castes population in whole of urban areas. In other areas of buffer zone their population is more than 10%.

Population of SC & ST in Buffer Zone

There are five Scheduled Castes communities inhabiting in the buffer zone. These are Chamar, Basor, Katia Kotwar and Mehra. Among them Chamar is the dominant group as they are inhabiting near about 120 villages in buffer zone. Mehra in 82 villages, Katia 65 villages, Kotwar in 44 villages and Basor only in 20 villages. Their total population spread in the entire buffer zone is 68836 of which Hoshangabad shares 50841 (73.9%), Chindwara 15036

(21.8%), Betul 2959 (45.34%). Again if the total population of Scheduled Castes of Hoshangabad, Betul and Chindwara is distributed amongst its segments or tehsils in buffer zone then the break-up of the population with percentage in these segments are as follows – Babai 4648 (9.14%); Bakhedi 10342 (20.34%); Block-Hoshangabad 659 (1.3%); Sohagpur 11643 (22.9%), Itrasi-Kesla 7454 (14.66%); Pipariya 16095 (31.66%); Shahpur 2469 (83.4%); Ghoradongri 490 (16.6%); Jamai 11989 (79.7%) and Tamia 3047 (20.3%). So, above mentioned facts clearly indicate that Scheduled Castes population are in sizable number and register their presence everywhere within the buffer zone.

Table 6.11
Population of Scheduled Castes in Buffer Zone

Tehsils of Districts	*Districts*			*Total*
	Betul	*Chindwara*	*Hoshangabad*	
Babai	00	00	4,648	4,648
Bankhedi	00	00	10,342	10,342
Block – HB	00	00	659	659
Sohagpur	00	00	11,643	11,643
Itarsi – KE	00	00	7,454	7,454
Pipariya	00	00	16,095	16,095
Ghoradongri	490	00	00	490
Shahpur	2,469	00	00	2,469
Jamai	00	11,987	00	11,987
Tamia	00	3,047	00	3,047
Total	**2,959**	**15,036**	**50,841**	**68,836**

(Source: Census of India, 2001)

Population of Scheduled Tribe in Buffer Zone

Four tribal groups form the entire Scheduled Tribe population of the buffer zone. They are Gond, Korku, Bharia and Mawasi. Among these, Gond and Korku together account for 90% of the tribal population of the buffer zone. Korku and Mawasi live in 107 villages (20.26%) of the buffer zone, but the majority of Mawasi population is living in the villages of Pachmarhi sanctuary and outside Pachmarhi sanctuary in Pipariya and Jamai tehsil. During field investigation the authors have been informed that Korku and Mawasi, the two tribes are clearly distinct one from the other and Mawasi maintain their own identity. Although, it has been observed by the authors that their aspects of social organization, religion and level of economic development is identical to Korku in many ways. In addition to that the habits of life of the two tribes i.e. Korku and Mawasi are much the same and in personel appearance they are not unlike each other. Hence, the census reports and Gazetteer also bracketed Mawasi in the order with Korku.

Table 6.12
Population of Scheduled Tribes in Buffer zone

Tehsils	*Districts*			*Grand Total*
	Betul	*Chindwara*	*Hoshangabad*	
Babai	00	00	5,802	5,802
Bankhedi	00	00	16,619	13,619
Block – HB	00	00	304	304
Sohagpur	00	00	14,753	14,753
Itarsi – KE	00	00	21,624	21,624
Pipariya	00	00	17,031	17,031
Ghoradongri	7,262	00	00	7,262
Shahpur	11,586	00	00	11,586
Jamai	00	28,885	00	28,885
Tamia	00	40,434	00	40,434
Grand Total	**18,848**	**69,319**	**73,133**	**1,61,300**

(Source: Census of India, 2001)

The Gond are the most numerous tribe living in 397 (75.18%) villages of the buffer zone. The Gond population living in the buffer zone are probably Raj Gond who perceived themselves as “Gond Thakur”, although, all of them may not be included under the category “Raj Gond”, sub group of Gond who are traditionally referred in the past and enjoyed higher status. Bharias living in 53 (9.84%) villages of buffer zone excluding Shahpur and Ghoradongri in Betul district. It is seen from the table that the total population of Scheduled Tribes living in buffer zone is 161300 (32.31%). So, out of 161300, Hoshangabad shares 73133 (45.33%), Chindwara 69319 (42.98%) and Betul shares the lowest population of Scheduled Tribe 18848 (11.68%).

If this figure is further calculated on the basis of its distribution in tehsils/segments in buffer zone then the break-up of the population with percentage are as follows – Babai 5802 (7.93%); Bankhedi 13619 (8.44%); Block-HB 304 (0.18%); Sohagpur 14753 (9.14%), Itrasi-KE 21624 (13.41%); Pipariya 17031 (10.55%); Shahpur 11586 (61.47%), Goradongri 7262 (38.53%); Jamai 28885 (41.67%); Tamia 40434 (58.33%). Now this breakup in minute detail gives us an image that Scheduled Tribe population is every where in buffer zone and bigger than the population of Scheduled Castes.

Education and Literacy

Literacy and level of education attained by the people of buffer zone was very poor in the past because since generations they lived in the forest and were thus deprived of the education. But now the situation is completely changed and the buffer zone is enjoying a good percentage of literacy rate. The existence of a number of primary schools in the villages of buffer zone and high schools and colleges in the neighbouring towns are mainly account for the educational progress in the area. Change in the attitude of the people in favour of education is also one of the reason for spread of education.

Table 6.13 gives a clear idea regarding literate and illiterate persons among male and female population of the buffer zone.

From the above mentioned table it appears that in Hoshangabad, Sohagpur has the highest literacy rate among male (70.08%) and Pipariya has the highest illiteracy rate (48.97%). Again Pipariya registers the highest literacy rate among female (39.79%) and Block-Hoshangabad the highest illiteracy rate (61.26%). In Chindwara Tamia has the highest literacy rate and Jamai the lowest (42.58%) but in case of female Jamia stands highest i.e. (41.37%). In Betul Shahpur has the highest literacy rate among both male (61.21%) and female (44.68%) but among illiterate male (42.09%) and female (76.25%), Ghoradongri has the highest percentage.

Civic Amenities

Drinking Water – Amenities like drinking water, road electricity are available to the people of buffer zone. Three major source types of drinking water i.e. Hand Pump, Tube Well and Well are found in the villages of buffer zone. People are utilizing

Table 6.13

Total Number of Literate and Illiterate in Buffer Zone

District	*Tehsil*	*Literate*					*Illiterate*				
		Persons	*Male*	*%*	*Female*	*%*	*Persons*	*Male*	*%*	*Female*	*%*
Hoshangabed	Pipariya	64456	33631	52.18	25653	39.79	39876	19531	48.97	22479	56.37
	Block – Hoshangabad	2502	1637	65.42	865	34.57	1771	686	38.73	1085	61.26
	Itarsi – Kesla	25358	15762	62.15	9596	37.84	23870	9879	41.38	13991	58.61
	Sohagpur	40413	28321	70.08	14092	34.86	33390	13792	41.30	19598	58.69
	Babai	22659	14126	62.34	8533	37.65	18133	7243	39.94	10890	60.05
	Bankhedi	23926	14879	62.18	9047	37.81	25251	10456	41.40	14795	58.59
Chindwara	Tamia	25000	16027	64.10	8985	35.94	31428	13158	41.86	18569	59.08
	Jamai	41020	24047	58.62	16973	41.37	34040	14497	42.58	19684	57.82
Betul	Shahpur	12558	7687	61.21	5611	44.68	11229	4452	39.64	6919	61.61
	Ghoradongri	10985	6542	59.55	4338	39.49	10940	4598	42.09	8342	76.25

(*Source*: Census of India, 2001)

all the three sources for drinking water. Hand Pumps and Tube Wells are provided by the govt. agencies like Statutary Panchayat but in most of the cases wells are owned by the individual household. Individual wells are within the premises of the house hold but Tube wells and Hand Pumps are in some cases near their premises and in some cases away from their premises.

In urban areas of buffer zone drinking water facility is available to the people through tap water, tube wells and hand pumps. Majority of the houses are connected with tap water. The people of the area are not using tank, pond, lake, river, canal, spring or any other sources as source of drinking water.

Electricity – The electricity is not available in all the villages of the buffer zone. The local people informed that only fifty percent of the villages of buffer zone have been electrified under rural electrification scheme by the govt. The household of these villages have been electrified and they are getting electricity, but at present the situation is worst as they are facing the problem of load shedding continuously for more than twenty hours. So, in absence of electricity the household of the villages of entire buffer zone are using kerosene and other oil as source of lighting. The urban areas of buffer zone are electrified but facing the problem of load shedding so they are also using kerosene and candle or emergency lamp, operated by battery as the source of iighting to their household.

Medical facility is also available in the areas of buffer zone through Primary Health Centre (PHC), Hospitals and Private Dispensaries. The urban areas of buffer zone have better medical facilities as a number of private doctors are practicing. In rural areas the medical facility is very poor and it is served through Primary Health Centre (PHC), but as informed by the local people these "PHC" are running without doctors and adequate medicines. At the time of emergency the villagers have to rush to the district hospitals.

The villages of buffer zone are served by pucca road and *kachha* road. The forest villages of buffer zone are hilly and forested so there roads are in bad shape. The villages which are served by kachha road are totally cut off during rainy season and the villages have to face a lot of problem. The overall condition of the roads in buffer zone are not up to the mark only the state highway connecting

Pachmarhi to district headquarter (Hoshangabad) is in good condition.

Conclusion and Summary

Pachmarhi once predominantly a cluster of isolated villages is now emerging as one of the important tourist place. There was a time when the tribals were the chief element of its population, but now people from different parts of Madhya Pradesh, in particular and India in general have formed the core of its present day population. This has brought about many demographic changes and their interference has tremendous impact on the whole of Pachmarhi region. As result many changes are taking place in the environment of Pachmarhi Ecological Components – land, water, forest, almosphere and habitat have never been so threatened earlier as at present. This state of affairs necessitated the urgency to protect the ecological components of Pachmarhi from further deterioration. So, accordingly on 03rd March 1999 Pachmarhi was included in the list of Biosphere Reserve and demarcated in two zones.

(i) Core Zone and (ii) Buffer Zone

The buffer zone is the area which adjoins or surrounds the core zone. The authors undertook the comprehensive study of the villages of Buffer Zone to examine in details the pattern of ethnic composition of the villages. The buffer zone is 4927.28 sq. km. It covers three districts of Madhya Pradesh namely Hoshangabad (2971.59 sq. km.), Chindwara (1386.56 sq. km.) and Betul (569.13 sq. km.). Thus, maximum area falls under Hoshangabad district. The forest area is as much as 62.47% of the total area while agriculture land is 29.38%. There are total of 528 villages of which 465 belong to revenue villages and rest 54 forest villages. The human habitation in the buffer zone of Pachmarhi biosphere reserve is distributed within these 528 villages and in nine urban areas of which only two Pachmarhi and Tawa Nagar Ranipura) fall within central part of biosphere and rest of them are located on periphery of the boundary. The buffer zone also includes Pachmarhi and Bori sanctuaries.

In the buffer zone as per 2001 census there is 499135 persons as against 407504 persons in 1991 census which shows an increase of 91631 persons over one decade period in the entire buffer zone. The population increase may be attributed to the extension in agriculture, irrigation facilities and urbanization. The male population of buffer zone is 260374 and female population is 238762, which reflects the majority of male population in entire buffer zone. Probably ecological condition and mode of living appear to be associated with increase in the male population. The population of buffer zone is composed of two major groups i.e. tribals and non-tribals. Bharia, Gond, Korku and Mawasi form the tribal groups whereas non-tribals belong to different constitutional category and religious groups. Repatriated Bengali population who are commonly known as "Bengali refugees", are also included in the population as they have been settled down by the government of India. Bulk of population of the ethnic groups, however, belong from the villages of the Buffer Zone.

When we give a cursory eye to the distribution of ethnic groups in the villages of buffer zone their distribution seems to be heterogeneous in nature. In forest villages of Pachmarhi and Bori sanctuaries Gond, Korku and Mawasi are in full majority and form the basis of greater human habitation. In revenue villages too they are in majority adding Bharia with other groups such as Brahman, Gadaria, Kurmi etc, but ethnic groups belonging to the Scheduled Castes category is completely absent here. The reason may be the problem of communication in these villages which are mostly hilly and forested. In Pipariya and Sohagpur the tribal population remains the same as mentioned above but the number of other ethnic groups are increasing considerably and Dhanuk, Gujjar, Kachhi, Nai, Purbia, Teli, Raghuvanshi, Muslim are found in majority in the villages of these tehsils. Here Katia, Mehra, Kotwar, who are categorized as scheduled castes also opened their account and five villages where there is no tribal population, their presence is observed by the authors. Similarly Brahman, Rajput, Kurmi, Kalar, Dhobi, Chamar, Ahir, Basor, Kerar, Kumhar are found in the villages of Kesla-Itrasi, Babai and Bankhedi tehsils. The villages vary widely in the number of ethnic groups living within their boundaries. Some are lived by three or four groups while there are others with as many as six or seven ethnic groups. As regards the tribal population is concerned, Gonds are widely distributed in these tehsils, Korku and Bharia are found in Kesla-Itrasi and Bhankhedi but these two groups are missing in Babai tehsil.

The ethnic groups of Shahpur and Ghoradongri don't show any distinctive features, rather draw a picture of similarity with other parts of buffer zone in Hoshangabad district. Addition of new population in the name of "Bengali refugee" in Ghoradongri from Maharashtra and Chhattisgarh is an important aspect of the area as it captures the attention of anybody and everybody whosoever visit this area. There are eight villages exclusively occupied by them and these villages may be termed as uniethnic villages. They are trying to assimilate themselves totally with their new natives but yet succeeded and the process is still going on. The numerically dominant group Gond lives in each and every village of the area but the case of Korku is somewhat unique as every second village of Shahpur is having one or more Korku families but they are absent along with Bharia and Mawasi in Ghoradongri.

Tamia and Jamia or Junnardev are the part of buffer zone from Chindwara district. Here, also the ethnic composition of the villages and ethnic groups are not apart from its counterpart in Betul and Hoshangabad but stands apart from other areas of buffer zone in the pattern of distribution of Scheduled Castes population. Twenty-seven villages of this area don't have any Scheduled Castes population. Chamar, Katia and Mehra are the Scheduled Castes population of this area. Among these Chamar and Mehra are traditionally associated with household industries and educationally Katia appears to be better off from Chamar and Mehra. So, exodus of these groups in urban areas may be one of the reason of their absence. Ahir, Bania, Brahman, Lohar, Rajput, Badhai and Teli have also made their appearance in the villages of Tamia and Jamai or Junnardeo. As regards the tribal population only Korku is missing in this area, other wise Gond, Bharia and Mawasi are found in a sizable number. Gonds are living in 122 villages by no more than 03 to 04 groups but out of this 122 villages, 21 villages are uniethnic i.e. Gonds only. Similarly, Mawasi have also made their appearance in 44 villages. In 07 villages they are living as a single group, in 08 villages only Gonds are living with them and in rest 29 villages they are living with other ethnic groups.

Bharia is living in 06 villages as a single group and in 04 villages with Gond and other groups.

There is not much difference in ethnic complexity between the urban and rural areas of the buffer zone, but in actual composition of ethnic groups, however there is a good deal of differences between the two tracts. While tribal groups like Bharia, Gond, Korku and Mawasi dominates the rural tract, the Chamar, Katia, Mehra dominate the urban area. The OBC groups like Aggarwal, Bania, Dhobi, Nai, all the religious groups and the Kayastha, Bishnoi concentrates more in urban areas.

The population of Scheduled Castes in the buffer zone is below one lakh, although, their population comprises 06 groups. They are scattered all over the area but their major concentration is in Pipariya where they are 31.66% of their total population. They live in common villages mixed with other groups, taking to agriculture and other allied occupations. They have also attained some cultural progress but still maintain a social distance from the clean castes.

The buffer zone has a rich tribal heritage. It is to be mentioned here that the groups listed in the buffer zone as tribals are those which have been recognized as Scheduled Tribe. Their population is bigger than the Scheduled Castes and comprises of 04 tribal groups. They constitute 32.31% of the total population of the buffer zone. They are found every where in the buffer zone but their major concentration is in Tamia where they are 58.33% of their total population. Bharia, Korku and Mawasi retain much of their tribal life and are substantial plough cultivators even then they have a subsistence economy, while Gonds are well off and have gradually accepted the Hindu view and way of life as they are representative of the old aristocracy of the area. Now they are calling themselves as "Thakur". This process has further enhanced the differences between them and other groups and created a new level of cultural development.

Now it is clear from the above mentioned accounts that in the villages of buffer zone several ethnic groups are living who still preserve their language, original culture, religion, their ethnic territory and a number of other features which make it possible to classify the people concerned as a separate ethnic unity.

References

(a) Articles

(1)	Barth, F.	1956	"Ecological Relationship of Ethnic Groups in Swat, North Pakistan" *American Anthropologist*, Vol. 58; No. 06
(2)	Frake, C. O.	1962	"Cultural Ecology of Ethnography" *American Anthropologist*, Vol. 64, No. 01, Part – 1
(3)	Netting, R. M.	1974	"Agrarian Ecology" *Annual Review of Anthropology*, Vol. 03
(4)	Vayda, A. P. and B. McCay	1975	"New Directions in Ecology and Ecological Anthropology" *Annual Review of Anthropology*, Vol. 04.

(b) Books

(1)	Donald L. Hardestry	1977	*Ecological Anthropology* John Wiley & Sons, Canada.
(2)	Gadgil, Madhav and Guha, Ramchandra	1995	*Ecology and Equity* Penguin Books India (P) Ltd., New Delhi.
(3)	Mohan, Madan	2000	*Ecology and Development* Rawat Publication, Jaipur
(4)	Reddy, K. Vishwanatha	2002	(ED) *Cultural Ecology of Indian Tribes* Raj Publications, Delhi.
(5)	Lewis, Michael	2003	*Inventing Global Ecology* Orient Longman Private Limited, New Delhi.

(c) Government Publications

(1)	Census of India	1961	*District Census Handbook* Betul District, Govt. of Madhya Pradesh, Bhopal.
(2)	Census of India	1961	*District Census Handbook Chindwara District* Govt of Madhya Pradesh, Bhopal.

(3)	Census of India	1961	*District Census Handbook Hoshngabad District* Govt. of Madhya Pradesh, Bhopal
(4)	Census of India	1971	*District Census of Handbook Betul District* Govt. of Madhya Pradesh, Bhopal
(5)	Census of India	1971	*District Census Handbook Chindwara District* Govt. of Madhya Pradesh, Bhopal
(6)	Census of India	1971	*District Census Handbook Hoshangabad District* Govt. of Madhya Pradesh, Bhopal
(7)	Census of India	1991	*District Census Handbook Hoshangabad District* Govt. of Madhya Pradesh, Bhopal

C.D's

(8)	Census of India	2001	*Primary Census Abstract Chhattisgarh & Madhya Pradesh* Registrar General of India, New Delhi

Reports

(9)	EPCO	2001	*Pachmarhi Biosphere Reserve* Vol. 01; No. 02, Bhopal
(10)	EPCO	2002	*Pachmarhi Biosphere Reserve* Vol. 01; No. 02, Bhopal
(11)	Das, R. R. and Singh, R. P.	2003	*Proceedings of National Seminar on Biodiversity* Environmental Planning and Co-ordination Organization, Bhopal

7

Demographic and Geographical Aspects in the Context of Pachmarhi Biosphere Reserve (M.P.)

R. B. THAKRE

Introduction

It becomes necessary to have some restrictions or limitations about development activities and population growth in Pachmarhi Biosphere Reserve to conserve its beauty, natural resources, biodiversity and ecosystem. Ecotourism is one of the best solution to this problem. Pachmarhi in the Mahadeo and Satpura hilly ranges become a spot of attraction to all types of tourist, specially religious one. The topography, which decides the nature of soil, vegetation and drainage system in particular climatic condition of Pachmarhi Biosphere Reserve (PBR) have vital role in population distribution. The block-wise population distribution in PBR shows the pressure on Natural resources in particular block. The due importance is given to land use pattern in relation to population distribution. Biogas and solar energy application should be encouraged in PBR. There is a threat from high cattle population in particular block of PBR it may convert some of the forestland in to barren land. Interpretation of various types of maps of PBR prepared by Envrionmental Planning and Coordination Organisation, Bhopal gives wholesome picture of physiography soil types wildlife and population distribution.

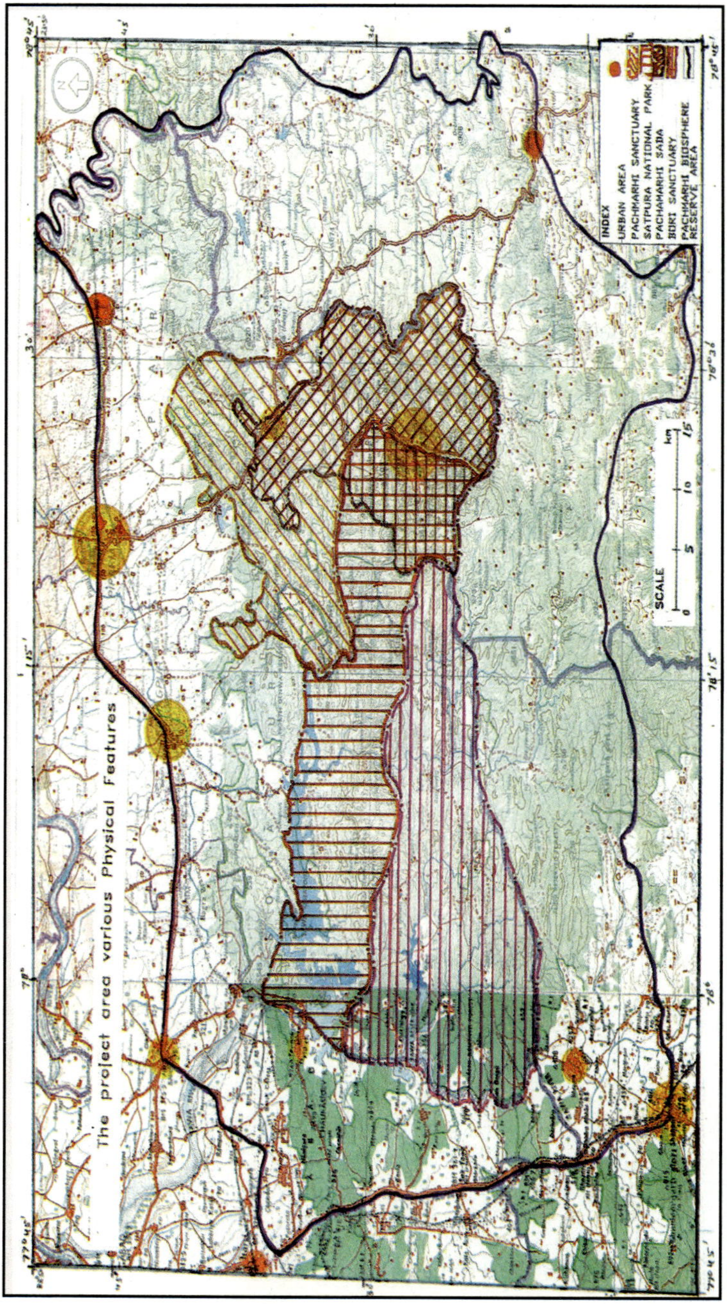
The project area various Physical Features
INDEX
URBAN AREA
PACHMARHI SANCTUARY
SATPURA NATIONAL PARK
PACHAMARHI SADA
BORI SANCTUARY
PACHMARHI BIOSPHERE RESERVE AREA
SCALE
km

Map 1

Map 2

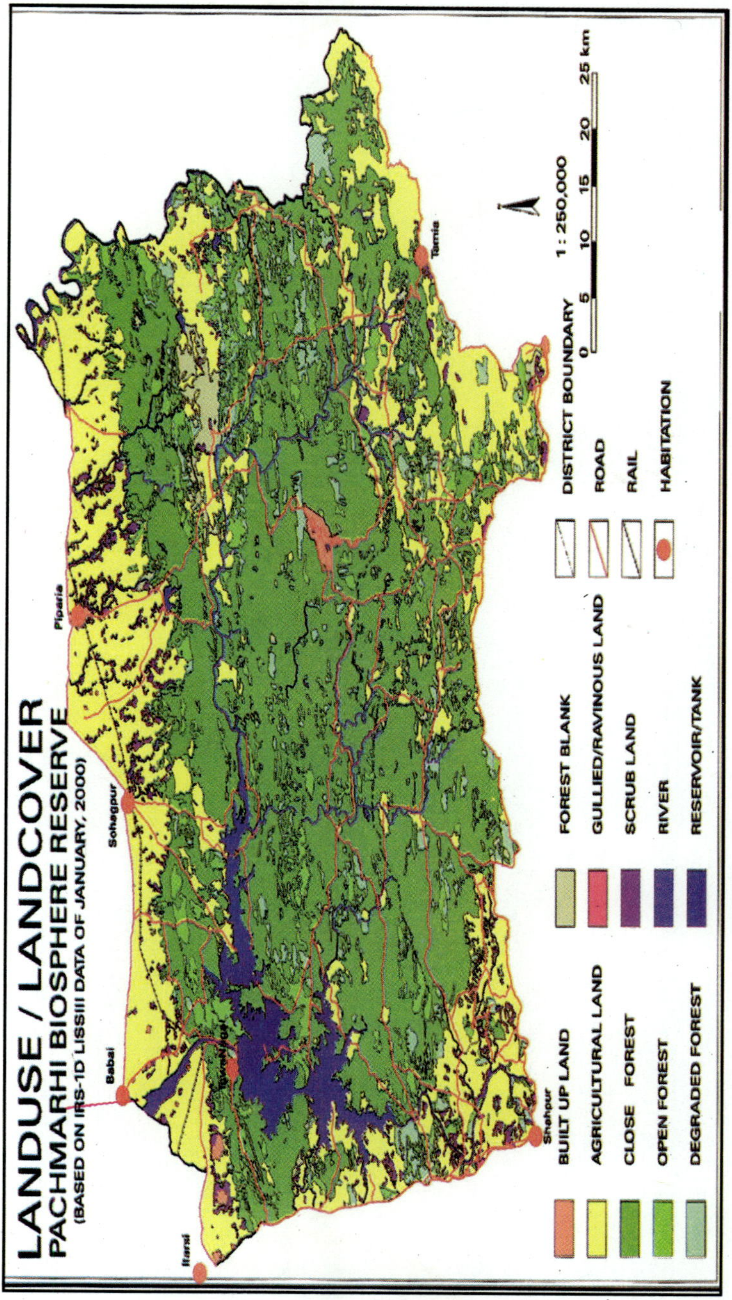
LANDUSE / LANDCOVER
PACHMARHI BIOSPHERE RESERVE
(BASED ON IRS-1D LISSIII DATA OF JANUARY, 2000)
Itarsi
Babai
Sohagpur
Piparia
Tamia
Shahpur
BUILT UP LAND
AGRICULTURAL LAND
CLOSE FOREST
OPEN FOREST
DEGRADED FOREST
FOREST BLANK
GULLIED/RAVINOUS LAND
SCRUB LAND
RIVER
RESERVOIR/TANK
DISTRICT BOUNDARY
ROAD
RAIL
HABITATION
1 : 250,000
0 5 10 15 20 25 km

Map 3

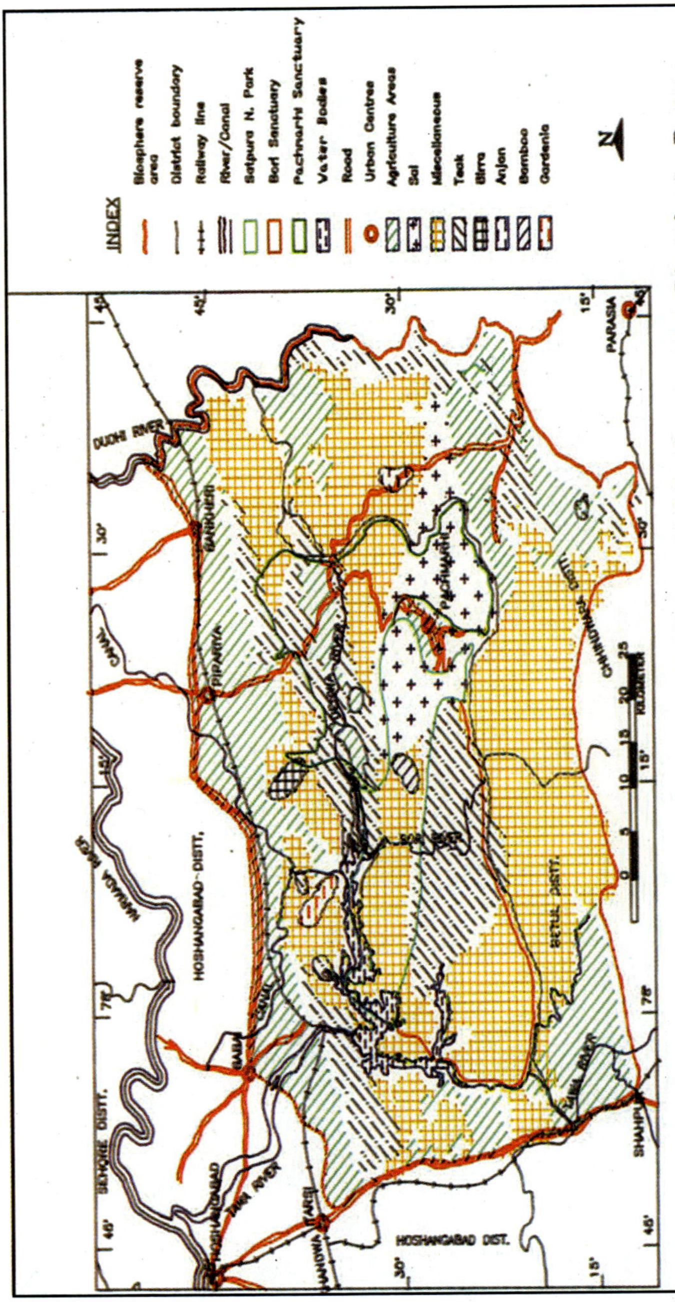

Forest types based on species composition of the Pachmarhi Biosphere Reserve

Map 4

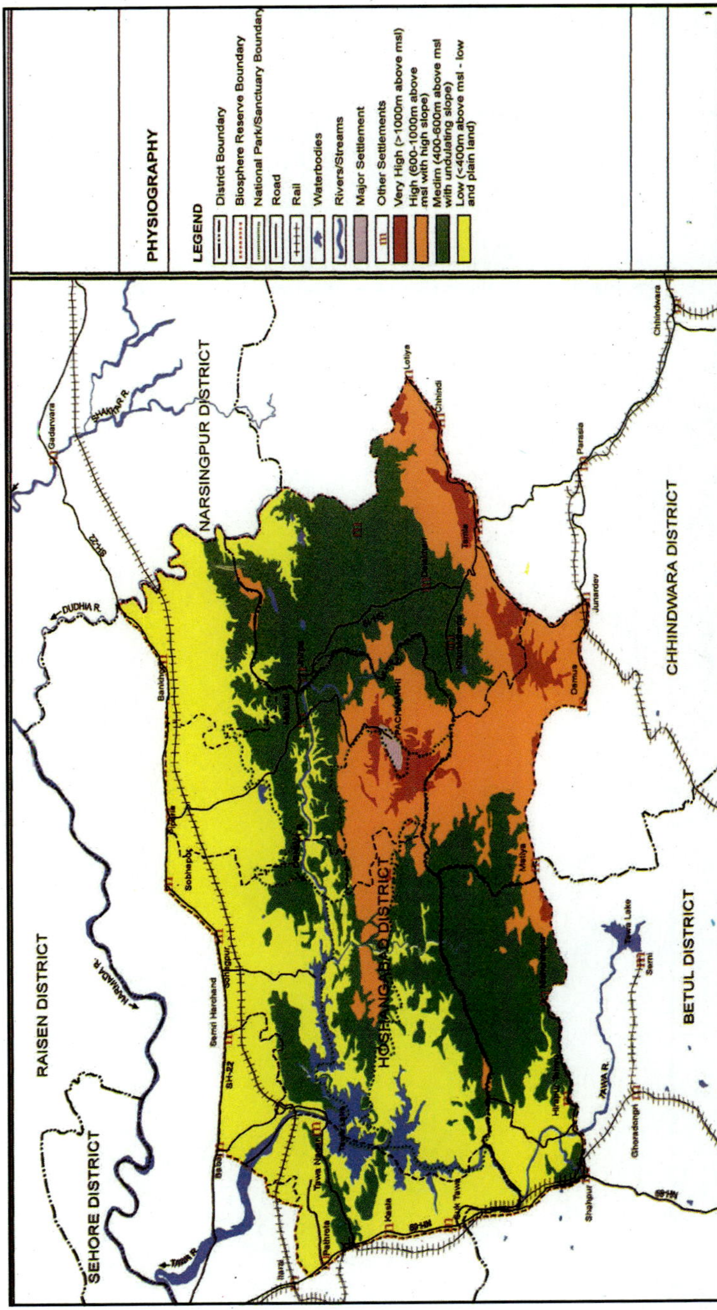
PHYSIOGRAPHY
LEGEND
District Boundary
Biosphere Reserve Boundary
National Park/Sanctuary Boundary
Road
Rail
Waterbodies
Rivers/Streams
Major Settlement
Other Settlements
Very High (>1000m above msl)
High (600-1000m above msl with high slope)
Medim (400-600m above msl with undulating slope)
Low (<400m above msl - low and plain land)
NARSINGPUR DISTRICT
RAISEN DISTRICT
SEHORE DISTRICT
HOSHANGABAD DISTRICT
CHHINDWARA DISTRICT
BETUL DISTRICT
PACHMARHI
NARMADA R.
TAWA R.
DUDHIA R.
Tawa Lake

Map 5

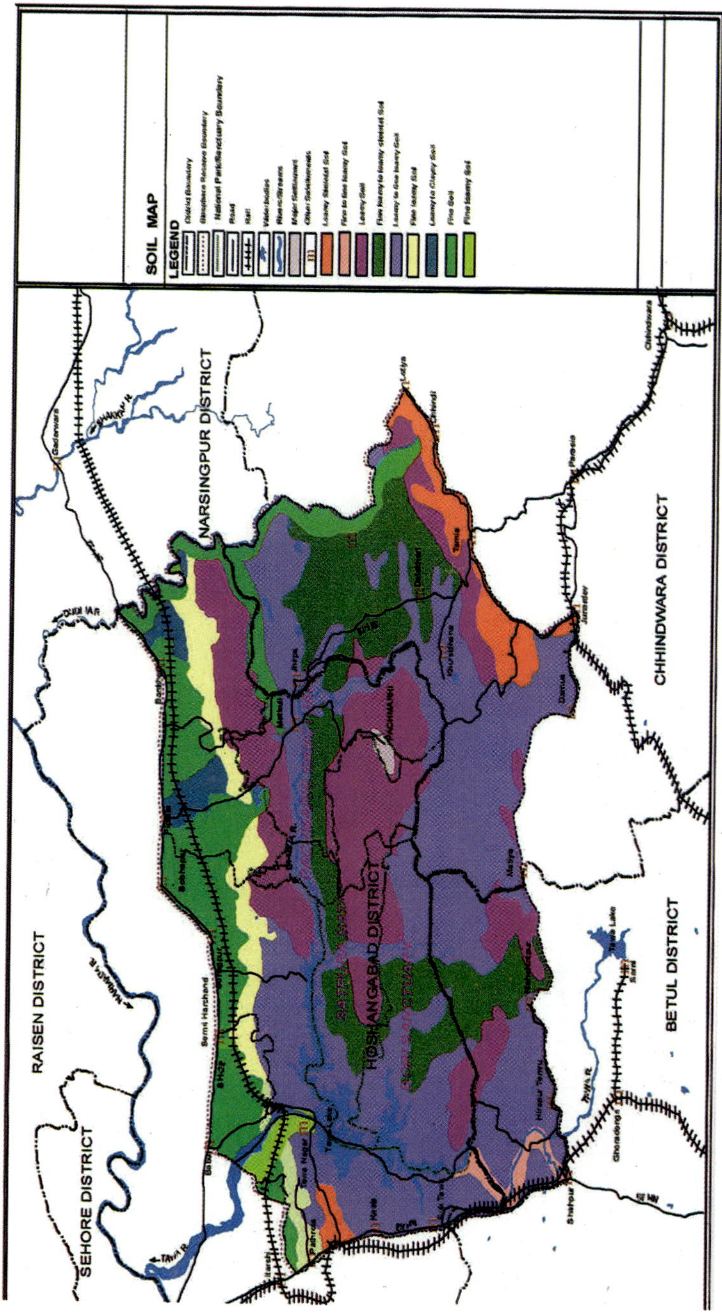
SOIL MAP
LEGEND
NARSINGPUR DISTRICT
RAISEN DISTRICT
SEHORE DISTRICT
HOSHANGABAD DISTRICT
CHHINDWARA DISTRICT
BETUL DISTRICT
PACHMARHI

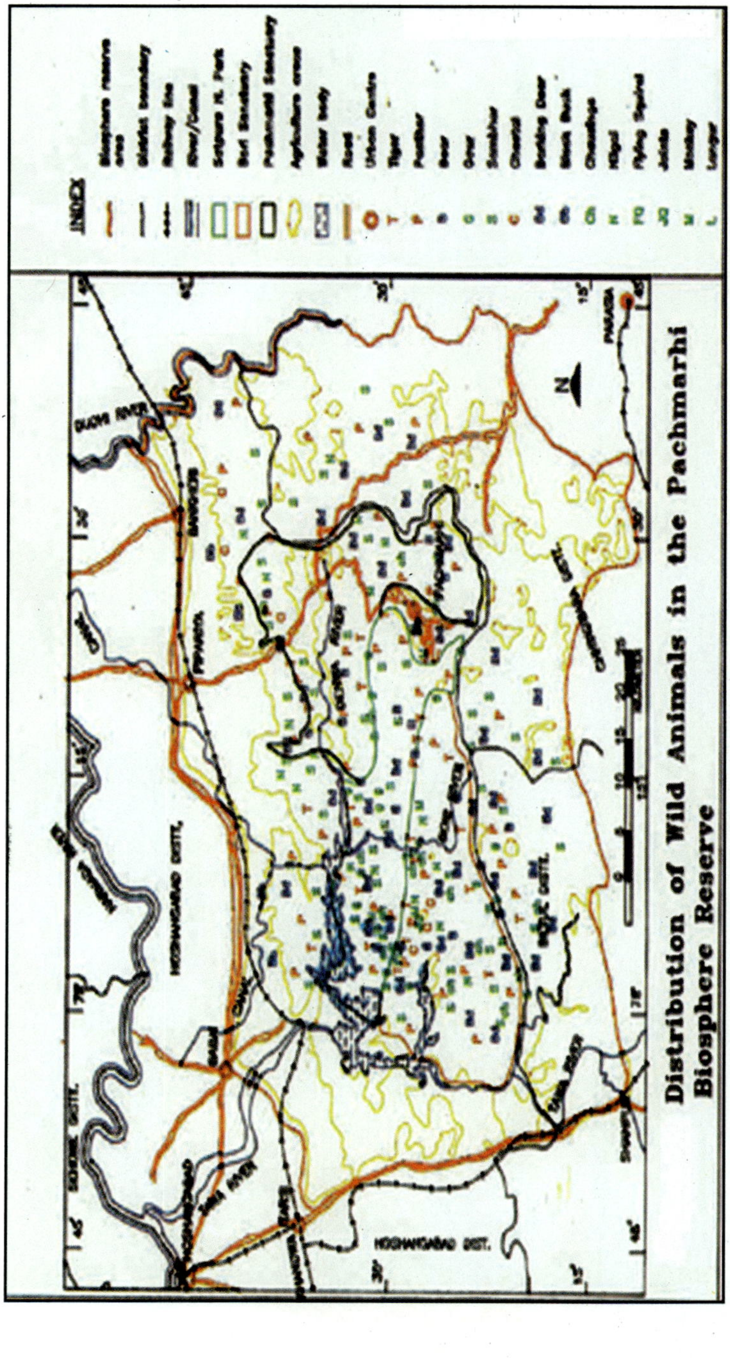

Map 6

Distribution of Wild Animals in the Pachmarhi Biosphere Reserve

Map 7

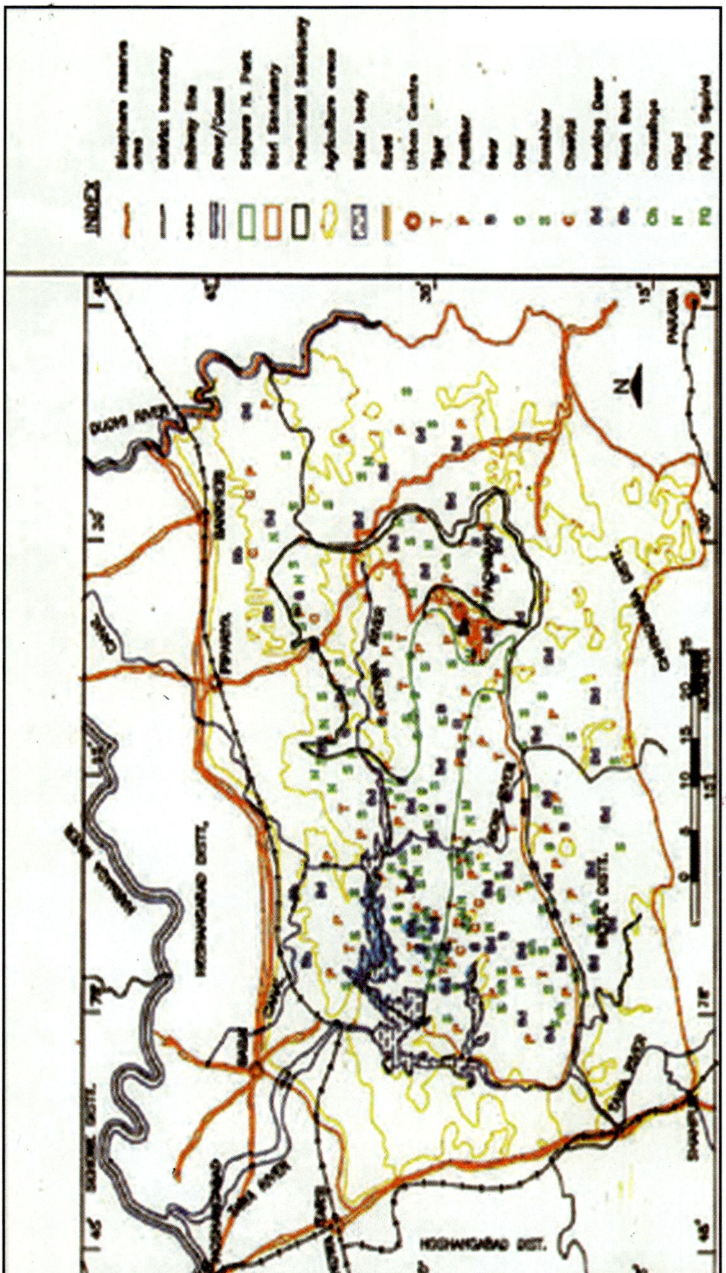

Map 8

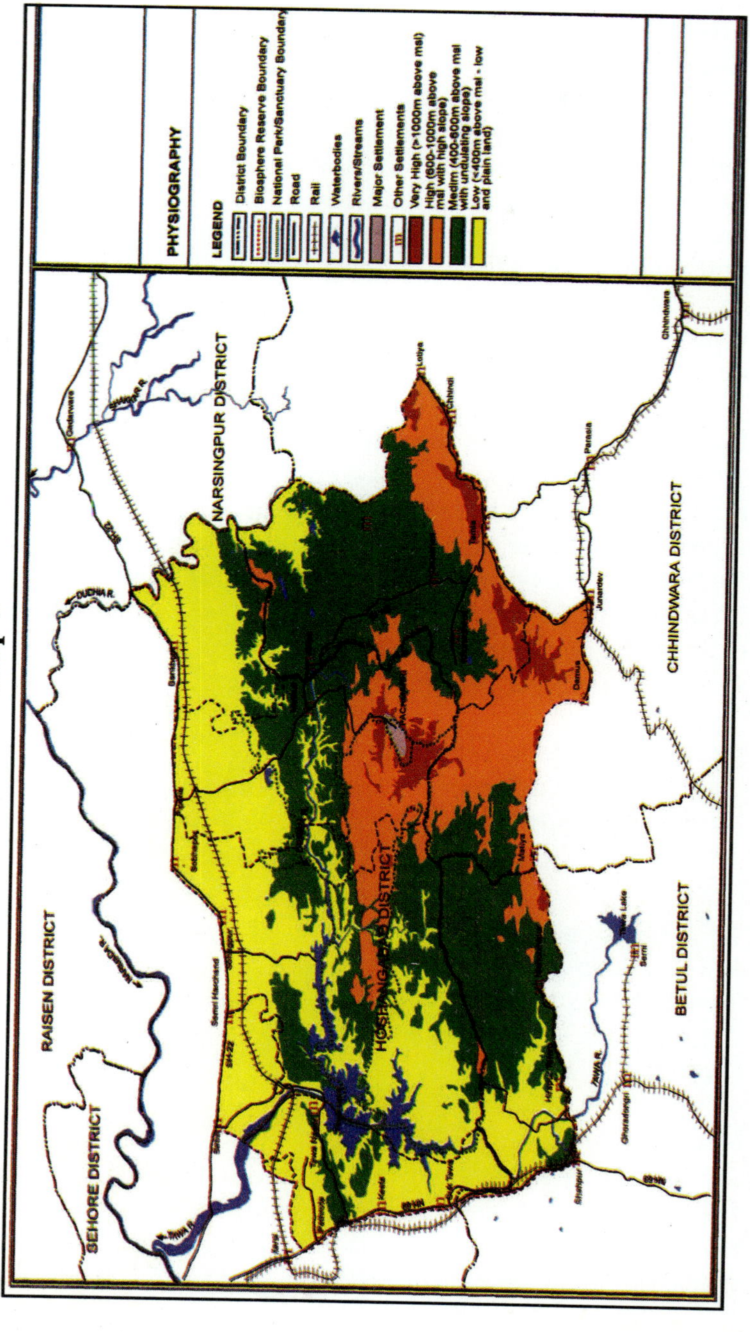
PHYSIOGRAPHY
LEGEND
District Boundary
Biosphere Reserve Boundary
National Park/Sanctuary Boundary
Road
Rail
Waterbodies
Rivers/Streams
Major Settlement
Other Settlements
Very High (>1000m above msl)
High (600-1000m above msl with high slope)
Medim (400-600m above msl with undulating slope)
Low (<400m above msl - low and plain land)
NARSINGPUR DISTRICT
RAISEN DISTRICT
SEHORE DISTRICT
CHHINDWARA DISTRICT
BETUL DISTRICT

ANALYSIS OF MAPS

Map 1 - Land Use/Land cover

In this map we see a complete boundary belt of 15 to 20kms. of agricultural land and settlement and a major area of closed forest in the central part of Pachmarhi Biosphere Reserve (PBR). Satpura National Park, Pachmarhi and Bori Sanctuary are predominantly a reserve area for maintaining natural resources and Biodiversity. The scattered patches of open forest and degraded forest are mainly surrounded by village settlement. It is clearly distinguished that the population pressure exerted from outer buffer zone towards central area of PBR.

Map 2 - Forest types based on the species composition.

This map shows Sal forest coverage mainly in Pachmarhi Sanctuary area. The coverage of Teak wood dominates in four area, that is Bori Sanctuary, Western part of Tawa Water reservoir, Denwa river basin on the Northern boundary of Pachmarhi Sanctuary and South Eastern belt of PBR. These two major species of forest Sal and Teak in PBR are surrounded by mixed forest, which has a maximum area in the PBR.

Map 3 - Map of Physiography

The central part of Pachmarhi and South Eastern part of PBR are highland hills and all major rivers originated from this area. This area is more than 600 metres high from sea level. This physical structures constitutes a radial drainage pattern. Other area of PBR have gentle slope towards its boundary. The central plateau and hilly area of PBR maintain healthy, cool and warm climate from March to May and then it becomes humid and cold up to October, then it becomes cold and dry climate from October to February.

Map 4- Soil map

Most of the central and other plateau area covered by loamy soil having fine grains of rocks and skeletals mixed with clays. Settlement and habitat area around the boundary belt of PBR specially in the river basin have fertile land of fine loamy soil. It has a limited area of irrigation near Babai and Hoshangabad block due to canal linkages. Other parts of PBR have limitations due to reserve area and highland topography.

Map - 5 Map of wildlife

The distribution of wildlife in PBR shows there is large-scale concentration of Tiger, Panther, Sambhar and beer in the Satpura National Park, Pachmarhi and Bori Sanctuary. The concentration of Chitah population is limited to central part of Bori Sanctuary and in the forest area of North Eastern PBR. The presence of Nilgai is in scattered form in the central area of PBR.

Map -6 Geology of PBR

There is a alluvial structure of about 10 km. which is around East, West belt on the Northern PBR boundary:

Then next to it towards South we find an east west belt about 10 to 15 km. rock structure of calcarious crystalline. Both these formation have gradual sedimentation through drainage system of Denwa, Dudhi and Tawa rivers.

Next to it, in the central part of PBR a large belt of East West direction are rock formations of Pachmarhi sandstones. This rock structure is broad on the Eastern side of PBR, covers above 15 to 20 kms. area whereas it becomes narrow on the Western side of PBR which comes about 5 to 10km wide. Pachmarhi Sandstone is spread in the large part of Denwa, Bori and Tawa river basin.

Next to it is 10 to 15 km. East, West belt in the Southern side of PBR. It is a rock formation of Bogra Denwa Conglomarate it is broad about 15 to 20 km on the Western side, whereas it becomes narrow up to 7 km on the Eastern side of PBR. In the central part of this formation of Bogra Denwa Conglomerate, state there is a small intervention of 5km wide intrusion of deccan trap in the central part of it. Deccan trap is basaltic in nature having hard structure of black stone and regur soil. A small patch of deccan trap flaws of 5 to 10 km concentrate in the South Eastern Corner of PBR. Adjacent to this corner a patch of 5 to 10 km wide formation is known as Bijori sandstone.

The geological map of PBR have clear-cut East West belts of different rock formation mentioned in the above analysis.

Map -7 Population distribution in PBR

Major concentration of population in PBR is along Itarsi Jabalpur railway line. The another reason of population concentration in this area is roadways from Pipariya to Hoshangabad. This belt of East West direction have low lying arable land and irrigation facilities, it comes under buffer zone of PBR so the restrictions of core zone Satpura National Park, Pachmarhi and Bori Sanctuary is not rigid in this area.

The second most dense populated area is near Junnardeo due to its fertile low level and having the communication network from Chindwara to Amla. The other area of PBR have moderate and scattered population. Western part of Satpura National Park and surrounding of Tawa Water reservoir both have Scanty population coverage.

Ecotourism is the best solution to minimize the population pressure on natural resources and maintaining the ecosystem. In the ecotourism there is scientific way of enjoying the nature's beauty without disturbing flora and fauna. Watching animals, enjoying photography, fishing, hunting, water sports, adventures etc, are some of the utilities of ecotourism in which we enjoy nature without disturbing and exploiting nature. It is the best-suited system for agriculture and forest based economy. Once the ecotourism is opened for all it will disturb the main objective of Biosphere Reserve that is preserving and conserving natural resources and biodiversity.

There is always a threat to local value culture and traditions if it opens for all. When the biosphere reserve opens for all the local tribals have to depend on external agencies and face some restriction to protect biosphere reserve.

There are 583 villages and urban agglomeration in PBR out of which 508 are revenue villages, 67 are forest villages and 8 are urban centers. The total area under Pachmarhi B. R. is 4981 km^2. the maximum area, about 59.54% are in Hoshangabad district, 29,20% are in Chhindwara district which is on the Eastern side of PBR is one of the most hilly, mountainous and plateau area. This hilly area is a part of Satpura, Mahadeo and Nalini group of mountain ranges. The general height of this area ranges from 600 to 1000 metres from sea level. Pachmarhi, Jamai and Tamia lies in this area of Southeastern part of PBR.

On the Northern side of PBR is surrounded by Raisen, Narsingpur and Sehore district of M. P. The Northern and Western belt of PBR is low lying area; the average height of this area is below 400 metres from sea level.

This area is drained by river Denwa and Bori River. This low-lying river basin area has most of the land under cultivation. This area has maximum settlement.

The large part of water reservoir is on the western part of PBR known as Tawa water reservoir. The central part of PBR is covered by Satpura National park having thin density of settlement due to many restrictions of Biosphere reserve. It is classified as closed forest protected from the exploitation of the people.

There are scattered patches of open forest away from the central part of PBR and those scattered patches of open forest are found mainly around the village settlement up to circular boundary area of PBR. The villages utilize the forest produce of this open forest around their villages for essential purposes.

There are deep gorges in the river bed of Denwa, Dudhia and Bori. Some of the waterfalls are the creation of soft and hard rock structure in the way of water flow of Nallah or river. The caves are formed in the basaltic rock mountain or in the riverbed. Jatashankar near Pachmarhi is one of the attractive caves. Denwa, Bori, Tawa and its tributaries are the part of Narmada river drainage system. Narmadas River, which flows from East to West direction, is 20 to 25 kilometre away from PBR border.

Forest: - The total area of closed forest in PBR is 85.3 per cent of total forest area, 8.2 per cent area is covered by open forest, 4.2 per cent is degraded forest existed near the open forest area which is a scrubland and utilized mainly for animal husbandry purpose by the villages. 2.3 per cent of forest area is blank forest. Teak and Sal is important species of forest in PBR. The forest in PBR is classified in 1) most deciduous 2) dry deciduous 3) central Indian subtropical hill forest.

Climatic Condition :- Hilly and mountainous region which is 1300 metres high from sea level and east west direction of Satpura ranges in PBR created cool and humid climate in the central part of PBR.

The low-lying area of river basin near Hoshangabad has hot summer and medium rainfall in rainy season. The average temperature of PBR is 28° centigrade. The maximum temperature in summer is 46° centigrade. The minimum temperature in winter is 10° centigrade. The average rainfall of PBR ranges from 600 to 2000 milimetres.

The landuse pattern of PBR :- The land use of PBR is based, on land sat imagery of April 1990 and Nov-Jan 1991-92. Out of the total land of PBR, which is 4987.38 kms, the forest cover maximum land that is 65.2 per cent of total area. Most of the area is closed and protected forest. The Satpura Tiger project in PBR, fulfill the objective of biosphere reserve. No doubt forest land is preserved due to PBR. Agriculture covers 27.7% of total land in PBR, which is the second most coverage in total land in PBR. Most of the agriculture land is all around the boundary belt of PBR, other areas have scattered patches around the village settlement. Water bodies cover 4.2 per cent of total land of PBR. Tawa water reservoir is major part of it, which is on the Western side of PBR. The land under wasteland is 2.5 per cent of total land. The wasteland is generally found between forestland and village settlement, it is near scrubland and between the undulating hilly tract. The percentage of built up land is only 0.5% of the total area of PBR. The most of the scrubland are on the Northern boundary belt of PBR , this scrubland is utilized as grazing land, by the inhabitants of that area. Large numbers of people graze their cattles on this grazing and scrubland.

Pipariya, Kesla and Bankhedi block have sizeable forest cover. The total area of forest in these block is 29803, 20665, and 15326 hectares respectively. The land under cultivation is maximum in Babai block about 71683 hectares is under cultivation followed by Sohagpur (65691 hectares), Bankhedi (59432 hectares), Pipariya (58363 hectares). The area under fallow land is more in Sohagpur block (2428 hectares) followed by Bankhedi (2226 hectares), Pipariya (2184 hectares). In Hoshangabd district, the area under food crop is maximum in Hoshangabad block (41670 hectares), followed by Babai (41091 hectars), Sohagpur (39309 hectares). Land under fruits and vegetables are more in Hoshangabad and Babai block. Land under fruits in Babai is 562 hectares. Hoshangabad is 502 hectares. Land under vegetable cultivation in Hoshangabad block is 1298 hectares, Babai 648 hectares. Land under chilli in Babai is 322 hectares, Bankhedi 241 hectares, Hoshangabad 138 hectares.

It is suggested that valuable water resources have to be utilized through micro irrigation works. For building micro irrigation work, local raw materials like stones, sands and soil can be utilized to build bunds. Local employment has scope in these works. Indigenous fruits and vegetables can grow in large numbers for maintaining good health of people and sound economy of the area. The local ponds and lakes can be utilized for large scale fishing purpose. It is also advisable to form Eco-development committee for the generation of employment and utilizing local resource in small scale and cottage industry. Awareness programmes are going to be arranged for conserving natural resources and maintaining eco system in PBR. There is a need to provide health facilities and protect inhabitants from water born and other diseases.

Geographical Distribution of Populations – (settlement)

Table 7.1 shows the population distribution in the blocks of Pachmarhi the Biosphere Reserve in 2001. The table shows the number of villages existed in each block, the total area of the block and percentage of population of each block with the total population of PBR is given.

The large concentration of population is found on the western part of PBR because, Delhi - Chennai railway route passes from Hoshangabad, Itarsi, Sohagpur, and Betul. This is low lying agriculture zone of PBR. Another large-scale settlement is on the East West railway line. On Westward, this railway line goes from Itarsi to Khandwa and on East Ward, small railway track goes from Ghoradongri to Sarni Thermal Power Station. This Sarni area is rich in coalmines and generates large scale employment in coalfield. Another railway line comes from the south eastern part of PBR that is from Parasia to Amla and connected with Delhi Chennai route. Parasia and Junnardeo are developing due to this line and utilizing the coal resources in large quantity.

The population is also concentrated in the important towns and villages on the Chhindwara, Parasia, Tamia, Pipariya and up to Hoshangabad road ways. This is important roadways of PBR

Table 7.1

Population Distribution of Pachmarhi Biosphere Reserve, 2001, Blockwise

Sr.No.	*Name of the Blocks*	*No of villages*	*Area in sq. km.*	*Population*	*Percentage of population*
(A)	**Hoshangabad District**				
1.	Kesla	52	681.68	49228	9.88
2.	Babai	24	581.67	40799	8.17
3.	Sohagpur	80	598.95	73803	14.80
4.	Pipariya	82	798.42	104332	20.90
5.	Bankhedi	49	628.10	49177	9.85
6.	Hoshanabad	5	529.37	4273	0.85
(B)	**Betul District**				
7.	Ghoradongri	39	679.58	21926	4.40
8.	Shahpur	24	527.70	23809	4.76
(C)	**Chhindwara District**				
9.	Tamia	112	1268.02	56728	11.36
10.	Junnardeo Jamai	61	1085.97	75060	15.03
		528	**7346.46**	**499135**	**100.00**

There is population pressure from all the four direction of PBR towards central part of Satpura National Park, where natural resource is plenty.

There are ten blocks of Hoshangabad, Betual and Chhindawara district of M.P. in PBR. Out of the total population of 499135 in PBR in 2001. The Pipariya block has highest population of 104332, which comes about 20.90 per cent of the total population of PBR. There are 82 villages of Pipariya block in PBR and this block has 798.42 sq.km area under its coverage. The density of population in Pipariya block is 91 persons per sq. km. Lower than the state average of 149 persons in 1991. The Pipariya block is located in the North Eastern part of PBR and Pipariya town of this block is well connected with railways and roadways. Pipariya town is a nodal center in North Eastern part of PBR. All types of commercial, agricultural and forest activities are controlled through this town. Small portion of Southern area of Pipariya block is covered by forest otherwise all other part is low-lying area. Primary occupation of this area is agriculture. The animal husbandry occupies the secondary and supplementary occupation of the people.

Sohagpur is the second most populated block of PBR. The total population of 80 villages of Sohagpur block is 73803 which accounts 14.80 per cent of the total population of the PBR. The total area under the Sohagpur block is 598.95 sq.km^2. and the density of population is 144 persons per sq. km. The southern part of Sohagpur block has deep forest cover. This forest is closed and protected forest in PBR. The Northern part of Sohagpur block is well connected with the communication and is in the developing stage.

Bankhedi block in the North Eastern side of PBR is the third largest populated block in PBR. There are 49 villages of this block in PBR whose total population are 49177, which accounts 9.85 per cent of the total population of PBR. The total area coverage of this block is 628.10 sq. km. and the density of population is 132 persons per sq. km. below the state average of 149 persons. Bankhedi is well connected with Bhopal- Jabalpur railways and roadways. It is a low-lying area. Below 400 m from sea level most of area of Bankhedi block is under cultivation only the Southern tip of Bankhedi block is covered by forest. The population pressure is exerted from North to South direction for explorating of natural resources.

Fourth largest populated block in PBR is Junnardeo (Jamai) in Chhindawara district. There are 61 villages of this block in PBR whose total population is 75060 and accounts 15.03 per cent of the total population of PBR. The total area of Junnardeo block is 1085.97 sq.km and the density of population is 160 persons per sq. km more than state average 149. Junnardeo is connected with Chhindwara Amla railway line and Chhindwara Pipariya / Pachmarhi roadways. Though Junnardeo is a highland plateau area most of the part of land is utilized for agriculture purpose. Nearly half of the land of Junnardeo block is covered by open and degraded forest.

Fifth largest populated block in PBR is Tamia, in Chhindawara district. There are 112 villages of Tamia block in PBR whose total population is 56728 in 2001. This accounts about 11.36 per cent population of PBR. The density of population of Tamia block is 160 persons per sq.km. The highland plateau and hilly ranges of Tamia block on the Southern side is more than 1000 metres high from sea level. The general slope of this block is on the North Western direction and the height of the land reduces up to 400 metre from sea level. Southern highland plateau of Tamia is a source of river Denwa which flows from Pachmarhi plateau and then flows towards west of PBR where it passes from Tawa water reservoir and meets with Narmada in the North of PBR. About 60 per cent of the total area of Tamia block is forested, majority of which is protected or closed forest and small part around villages are open forest or degraded forest. Tamia town is well connected with Chhindawara Hoshangabad or Pachmarhi roadways. Teakwood is a main species of forest in Tamia block. The tribal people of Tamia felt the restriction of closed forest in day-to-day life; ultimately they have to depend on open or degraded forest.

Kesla is the sixth largest populated block of Hoshangabad district in PBR. There are 52 villages of this block in PBR, The total population of these villages are 49228 which account 9.88 per cent of total population of PBR. Total area of Kesla block is 681.68 km[22] and density of population is 119 persons per sq.km.

Kesla is on the extreme border area on the western side of PBR. Maximum area of Kesla block is low-lying area below 400 metres from sea level and fully utilized for cultivation and animal husbandry. Kesla town is connected with the National Highway no. 69, which runs from North to South. The population pressure is from west to east direction towards the central portion of PBR.

The seventh most populated block in PBR is Babai of Hoshangabad district. 24 villages of Babai block is in PBR the total population of these villages is 40799 which accounts 8.17per cent of the total population of PBR. The total area under Babai block is 581.67 sq. km. and the density of population is 145 persons per sq. km.

The area of Babai block is low lying area below 400-metres height from sea level. Most of the land is under cultivation, only small portion of South Eastern portion of Babai block is covered by open forest. Babai is connected with Hoshangabad Pipariya Chhindawara roadways. River canal on the Bori river which flows from west to east and then meets with Narmada River is important source of irrigation in this area. The population pressure of Babai block exerted to Southern direction towards central area of PBR.

The eighth populated block in PBR is Shahapur of Betul district. There are 24 villages of Shahpur block in PBR. The total population of these villages are 23809, which accounts 4.76 per cent population of PBR. The total area of Shahpur block is 527.70 sq. km. and the density of population is 137 persons per sq. km. Shahpur is an important center of this block on the Bhopal Nagpur National Highway no. 69. Most of the land is under cultivation only small patches of open forest is on the North Eastern side of Shahpur block. These patches of open forest are covered by scrub and fallow land.

The Gharadongri block of Betul district have ninth position in population, size in PBR. There are 39 villages of Ghoradongri block in PBR whose total population is 21926. This population accounts to 4.40 percentage of population of PBR. The total area of Ghoradongri block is 679.58 sq. km. and the density of population is 143 persons per sq. km. Ghoradongri is a town on Delhi - Chennai railway line and an important Centre on South Western corner of PBR. The average height of this block ranges from 400 to 600

Table – 7.2

Density of Population and SC-ST Population in PBR-1991

Sr. No.	*Name of Dist/block*	*S.C. Population*	*% of S.C. Population*	*S.T. Population*	*% of S.T. Population*	*Density of Population*
(A)	**Hoshangabad Distt.**					
1	Kesla Block	15040	15.48	31884	41.18	119
2	Hoshangabad Block	13755	16.97	4621	5.70	149
3	Babai Block	11734	13.68	8300	9.68	145
4	Sohapur	15341	17.49	10035	16.03	144
5	Pipariya	14904	18.51	14612	19.40	91
6	Bankhedi	17352	20.85	14561	17.49	132
(B)	**Chhidawara District**					
7	Tamia block	4408	5.78	59130	77.61	60
8	Jamai block	27832	16.01	84263	48.48	160
(C)	**Betul District**					
9	Ghoradongri Block	7184	7.36	54001	55.34	143
10	Shahpur	6035	8.30	44338	61.02	137

meters from sea level. The general slope of this block, are from South East to North West direction. Most of the areas of Ghoradongri block are cultivable, only the North Eastern corner of Ghoradongri is covered by closed and open forest. There is population pressure from southwestern direction of Ghoradongri block towards Bori sanctuary on the Eastern side of PBR.

Hoshangabad block has five villages in PBR. Total population of these villages is 4273 which accounts 0.85 percentage of total population of PBR.

The distribution of population of PBR is shown by dot method in Map 7.

Tribal population is dominated mainly in Kesla block of Hoshangabad district, where 41.18 per cent population is tribal. Tamia and Junnardeo block of Chhindwara district have 77.61 percentage and 48.48 per cent tribal population respectively. Ghoradongri and Shahpur block of Betul district have also large number of tribal population that is 55.34 and 61.02 per cent of total population respectively. Table 7.2 shows the density of population sq. km. along with Scheduled Caste and Schedule Tribe population in each block of Pachmarhi Biosphere reserve in 1991.

Ethnic Background – Most of the villages in Buffer zone of PBR have Gond, Mawasi and Katiya community in large numbers. The percentage of this community ranges from 40 per cent to 70 per cent of the total population of the respective villages. The percentage vary from village to village. Other communities who have sizable number of existence are Harijan, Gawali, Lohar, Mehara, Rajput, Chamar, Pawar, Vishwakarma, Sahu, Chipa, Ahir, Bhoyar, Yadav, Vai, Dhimar, Brahmin and Baniya.

Conclusion

Pachmarhi Biosphere Reserves have plenty of spots having attractive scenic beauty; so there is lot of scope for developing the tourism industry in PBR. There is some restriction on developing tourism infrastructure in PBR area.

(i) There is great need to encourage the use of biogas and solar energy to conserve the natural resources of PBR.

(ii) There is a lot of pressure on natural resources in the PBR due to population growth in the region. There is a need to control the population explosion.

(iii) There is a serious problems regarding the disposal of water and waste material from the towns and villages in PBR.

(iv) The high cattle population is another pressure area in preserving and maintaining the forest resources of PBR.

(v) There is a need to prepare an appropriate policy to maintain and utilize the precious indigenous medicines available in PBR.

References

1. Fr. S. John Britto, Human Population and Biological diversity. In the Proceeding of National Seminar on Biodiversity, Conservation and Management. With special emphasis on Biosphere Reserve 24-27 November 2000. edited by Prof. R. R. Das and R. P. Singh, Published by Epco. Bhopal (M.P.)
2. Census of India 1991, Madhya Pradesh series 13. Primary Census Abstract General Population. Part –II B (1), Published by Director of Census Operation M. P. Bhopal.

List of Contributors

Sadan Sengupta (Retd.) Superintendent Anthropologist (Cultural), Eastern Regional Centre, Anthropological Survey of India, Kolkata.

D.N. Pandey, Human Ecologist, Central Regional Centre, Anthropological Survey of India, Nagpur.

A. Roy, Research Associate (Cultural), Central Regional Centre, Anthropological Survey of India, Nagpur.

Samit Ghoshal, Anthropologist (Cultural), Central Regional Centre, Anthropological Survey of India, Nagpur.

D. K. Samanta, (Retd.) Assistant Anthropologist (Cultural), Western Regional Centre, Anthropological Survey of India, Udaipur.

Surajit Kumar Pal, Research Associate (Cultural), Central Regional Centre, Anthropological Survey of India, Nagpur.

Tarnal Maity, Research Associate (Cultural), Central Regional Centre, Anthropological Survey of India, Nagpur.

Mukesh Sahay, Research Associate (Cultural), Central Regional Centre, Anthropological Survey of India, Nagpur.

Sheela Dasgupta, Research Associate (Cultural), Central Regional Centre, Anthropological Survey of India, Nagpur.

R.B. Thakre, Cartographer, Central Regional Centre, Anthropological Survey of India, Nagpur.

Index